NUCLEAR SAFETY HAS NO BORDERS:

A HISTORY OF

THE WORLD ASSOCIATION
OF NUCLEAR OPERATORS

PHILIP L CANTELON

This book is dedicated to the highly professional men and women of WANO who operate, maintain, and support nuclear power plants worldwide, and who understand that 'good enough' is the enemy of excellence

This edition was first published in 2016 by
World Association of Nuclear Operators (WANO)

Publishing Director
Rick Haley / Joel Bohlmann

Editor
Claire Newell

President
Seok Cho

Chief Executive
Peter Prozesky

Head Office
WANO London,
Level 35
25 Canada Square
London
E14 5LQ
United Kingdom
T: +44 (0)20 7478 9200
E: communications@wano.org
www.wano.info

Creative Design and Production
Graphic Evidence Ltd
www.graphicevidence.co.uk

ISBN: 978-1-5262-0483-7

Printed and bound by CPI Group(UK) Ltd, Croydon CR0 4YY

TABLE OF CONTENTS

CONTENTS

PREFACE

Nikolai Lukonin stood under the glare of stage lights in a large dining room in the State Palace of the Kremlin. Dressed in a plain grey suit and grey tie with his Hero of Socialist Labour and other medals proudly hanging from his chest, he looked very much like the former Soviet Minister for Atomic Energy he was – a bit out of place in modern Russia. But on this day Lukonin was a guest of honour at the dinner. He was the historical bookend, present at the creation of the World Association of Nuclear Operators (WANO) in 1989 and honoured nearly a quarter of a century later for his crucial role in bringing east and west together to form a do-it-yourself industry organisation that would improve the operations and safety of commercial nuclear power plants.

The 12th Biennial General Meeting (BGM) WANO held in Moscow in May 2013 was, in part, the beginning of a celebration of the organisation's first quarter of a century. The loose association of nuclear utilities forming an industry forum to share operating experience and technical counsel in order to self-regulate and improve the safety, reliability and operations of their plants had evolved into an integrated, confident global industry organisation determined to overcome the challenges of the past and to meet the changing landscape of commercial nuclear power's future.

There was a fitting symbolism in holding the 12th BGM in Moscow. WANO had been born in the very same building in 1989. Over that time, both the venue and the organisation had greatly changed. The Sovincentr, host to the inaugural meeting, had been transformed into the Crowne Plaza Moscow World Trade Centre. Although the structure still stood on the bank of the Moskva River, those who attended both meetings could not find any remaining interior landmarks of the former building in the new one. The geopolitical and nuclear landscapes had also changed dramatically over those 25 years. The Soviet Union had become the Russian Federation, and the country had become a major exporter of nuclear power plants. The US, Europe and Japan – the leading nations in the commercial nuclear power industry – had seen the centre of activity shift to Asia. WANO, once a fledgling association with a high risk of failing, had become a respected and necessary part of global nuclear safety.

Three Mile Island, Chernobyl, Fukushima – these were the defining events in the history of commercial nuclear power. Each incident reinforced the lesson that all nuclear power plants were "hostages of each other", as William S Lee III, president of Duke Power and a driving force behind the creation of the Institute of Nuclear Power Operations (INPO), stated after Three Mile Island. A devastating accident at one plant had enormous repercussions on all. Yet Three Mile Island had little impact globally. Such an accident might occur in the US, international operators thought, but they were better than their American counterparts at running nuclear plants and adopted an "it can't happen here" attitude. Chernobyl changed that. Again, 22 years later, Fukushima dispelled the complacency that had settled in from the successes of the intervening years. In each instance the industry responded, first in the US with the creation of INPO in 1979, then with a global organisation, WANO, in 1989, and lastly with a restructured and refocused 'One WANO' by 2015.

However, to emphasise solely these major events would be to overlook the steady achievements of these exceptional voluntary industry organisations. INPO had established significant programmes to improve nuclear safety in North America. Yet any American who believed that the actions taken by nuclear utilities in the US after Three Mile Island could be imported wholesale into the global commercial nuclear power market was badly mistaken. The international ownership model – governments owned or controlled utilities in many countries – was different and more complex than in the US. In addition, cultural and linguistic differences defied simple solutions. The WANO compromise of creating four regional centres was brilliant, ensuring that all could participate within their cultural norms. The arrangement enabled and assured the organisation's survival during its formative years. Although the structure engendered friction, relations were never fractured. However, broad acceptance came slowly. Pedro Figueiredo of Brazil, who joined the Paris Centre Governing Board in 1991 and received a WANO Nuclear Excellence Award in 2003, recalled that "in the beginning, the majority of WANO members did not know what the main goals of WANO were other than safety." In addition, early peer reviews were met with considerable scepticism. Today, he observed, "there is a standard of excellence that is measured through peer reviews globally," and WANO had become an integral part of the nuclear operating experience.[1]

Figueiredo was correct in identifying peer reviews as being critical to WANO's success. Peer reviews were based on best practices; they were constructive, not chiding. Experienced reviewers recognised that WANO could not inoculate against error, but peer reviews could strive to anticipate and prevent errors caused by complacency, lack of resources or a slack safety culture.

Importantly, the "peer" of peer reviews held two meanings, both central to achieving excellence in nuclear operations and safety. One definition was a person of the same rank, ability and quality – an equal. These were the individuals who would inspect, analyse and evaluate the workings of a nuclear power plant. How they accomplished this was the second definition of peer – to observe, to look closely and searchingly for strengths and weaknesses, or areas for improvement, in the operations and operating culture of a nuclear power plant. Eventually, the concept of peers peering became a cornerstone of global nuclear safety.

Those achievements did not come easily. In more than a quarter of a century, WANO's history has passed through four stages of evolution. The first, from 1989–1996, was a direct result of Chernobyl and encompassed the initial organisation of the world's nuclear power operators into a self-policing confederation with the aim of improving the safety and performance of nuclear power stations. The confederation model established in the WANO Charter – four autonomous regional centres in North America, Western Europe, Eastern Europe and Asia, linked by a coordinating centre in London – was a brilliant and necessary compromise to ensure international acceptance and commitment. While adopting several programmes based on INPO's decade of industry experience in the aftermath of Three Mile Island, the major accomplishment during this period was WANO's success in opening communications between operators in the former Soviet Union and those in the West.

Thereafter, WANO turned to the broader safety concerns and practices of all its members. The second phase, from the mid-1990s to the early 2000s, was a period of consolidation and the reaffirmation of the basic core of WANO programmes – peer reviews, operating experience, technical assistance, and workshops and seminars – that emphasised personnel exchanges and information sharing. The third phase, from 2002–2009, consisted of an attempt to repair perceived shortcomings in the operations of the association stemming from the cultural differences of the four regions. An attempt to meld the four WANOs through a realignment of the governing structure, with an aim to centralise the power of the coordinating centre without eviscerating the autonomy of the regional centres, proved to be unsuccessful. The failure of the third phase to achieve those goals led to the fourth period, beginning in 2009 with a major overhaul of the governance structure to create greater involvement, accountability and consistency among the membership. As those governance changes were occurring, Fukushima caused WANO to rethink the adequacy of its activities in a changing nuclear power environment. WANO's response accentuated the importance and the urgency of revising, updating and completing the additional programmatic and operational changes recommended by the Post-Fukushima Commission to create a 'One WANO' from the four.

WANO's remarkable story holds compelling lessons for all global industries that must manage risk, overcome inherent limitations and pursue continuous improvement. WANO is an industry association, unique in that it puts purpose above profit and champions operational safety for the benefit of all rather than one. It is a story of an organisation dealing with the imperatives of the moment while planning for the future. Management experts define culture as the shared psychology in a working environment that sets habits and defines an organisation's identity. Leaders create and manage culture, according to Edgar Schein, a former MIT professor of management and author of *Organizational Culture and Leadership*. WANO leadership sought to expose shortcomings in what it viewed as the WANO safety culture, largely shaped by the safety standards established by Admiral Hyman Rickover and the US nuclear navy in the 1950s and '60s. But each region cherished its autonomy and was opposed to much oversight from a central authority. As a result, WANO's general definitions of a safety culture or what a peer review should be did not always mesh with those of the regional centres. Significant gaps, or discrepancies, existed between the stated values of the organisation and their application among the regions. The history of WANO is the effort of the world's nuclear power operators to bring those values and activities closer together and to inculcate a common culture of continuing improvement and excellence in nuclear safety throughout the industry.[2]

****

For many years of its history, WANO operated under the radar, outside of public awareness. However, more recently there has been a concerted effort by the WANO Governing Board and the membership to more clearly explain what it does to a broader audience. This history is part of that initiative. In line with this policy, the Governing Board allowed me access to its meeting minutes and any additional documentation provided by the regional centres. Although I tried to be inclusive and balanced in writing this history, the perspectives of regional governing boards are often absent. In such instances, I have relied on the oral histories of individuals involved in regional discussions.

Many people contributed to this book, and I am deeply indebted to them. Zack T Pate and Laurent Stricker initially recognised the value of preserving and recording the history of WANO and, with George Felgate, Ken Ellis, George Hutcherson, Jacques Régaldo, Hajimu Maeda and Vladimir Asmolov, offered encouragement, comments and valuable information throughout the project. I greatly appreciate the insight of James O Ellis, a former member of the WANO Governing Board and President and CEO of INPO during the Fukushima crisis, who provided a sharp perspective on INPO/WANO relations during that period. Any author would be fortunate to have these men in his corner. I am grateful for the help of Debbie Sims of Atlanta Centre, the unofficial archivist of WANO, who provided many of the written records and published materials from WANO's past. The history could not have been told without her cheerful, dependable and prompt assistance. Jade Knowles, Pavel Choudhury and Claire Newell of the London Office provided valuable support and documentation. I also thank Vera Lukyanova of Moscow Centre, who smoothed all the oral history arrangements in Russia, and Anatoly Kirichenko of Rosenergoatom. In addition, the project could not have been done without the administrative assistance and suggestions of several members of the London Office staff, George Felgate, Rick Haley and Joel Bohlmann, who ably shepherded the project from beginning to end. I also want to acknowledge the assistance of the centre directors during the project, Dave Farr of Atlanta Centre, Mikhail Chudakov of Moscow Centre, Ignacio Araluce of Paris Centre and Hal Shirayanagi of Tokyo Centre. Closer to home, Gail Mathews and Vanessa Lide of History Associates provided careful reviews and astute comments on the manuscript, and Camille Regis coordinated the transcriptions of the oral histories. My wife, Eileen McGuckian, travelled the entire WANO journey with me and provided both editorial and moral support throughout. Finally, I want to thank all the oral history participants who shared their memories and perspectives on WANO's past. A list of those individuals can be found in the appendix. The WANO story could not have emerged without their personal contributions.

Philip L. Cantelon
Rockville, Maryland

CHAPTER ONE

FORGING A **GLOBAL SAFETY NET**

The large conference hall at the Sovincentr in Moscow buzzed with excitement as more than 300 delegates moved down the aisles and filled the seats in the pale blue auditorium. In the front of the room a simple stage was set with a small table, three chairs, a blotter, several fountain pens and stack of papers. A Soviet flag and a large conference logo hung on a curtain in the background. Six translator booths lined the back of the auditorium. In this ordinary setting, something extraordinary was happening: the ratification of the World Association of Nuclear Operators (WANO), an industry agreement to create a non-governmental international organisation with the goal of improving the operation and safety of nuclear power plants worldwide. The meeting was, according to an industry publication, the beginning of a "nuclear glasnost".[1]

The formal signing ceremony to create WANO began at 14:00 on Monday 15 May 1989. On stage stood the co-chairmen of the conference, Nikolai F Lukonin, the USSR Minister for Atomic Energy and President of the Soviet Organising Committee, and Lord Walter Marshall of Goring, the Chairman of Britain's largest utility, the Central Electricity Generating Board (CEGB) and Chair of the WANO Steering Committee, which had drafted the charter of the new organisation. The two men were unlikely partners. Lukonin was an electrical engineer, former Director of the Leningrad nuclear power plant and devoted Communist Party technocrat who rose through the ranks to become the country's first Minister for Atomic Energy; Marshall was a world-renowned and award-winning theoretical physicist, a former Chairman of the UK Atomic Energy Authority (AEA), the head of the CEGB and a product of the capitalist system. Conservative Prime Minister Margaret Thatcher had awarded Marshall a peerage for his ability to keep the country's lights on during a protracted coal miners' strike in 1984–1985. Of medium build, Lukonin favoured plain grey suits in the tradition of Soviet officials. But whatever he wore, a shock of unruly hair gave him a naturally rumpled look. Marshall's tailored suits covered a tall, heavy-set frame. His reading glasses, perched halfway down his nose, gave the impression, fair or not, that he was always looking down on something or someone. He was an imposing figure. As one associate observed, "he could fill up a room," both with his stature and his personality. Off to the side stood the official photographer, poised to capture each member as he signed the WANO Charter. Although the ceremony was only one part of this inaugural meeting, it was highly significant that in signing the Charter each of the world's nuclear leaders was making a commitment in front of his peers to work toward the improvement of nuclear power safety and performance – a vow, one said, "to work harmoniously to improve the quality of operation of nuclear power plants". The disaster at Chernobyl in the spring of 1986 had brought the conferees together with the realisation that global cooperation was necessary for advancement of nuclear safety and the future of nuclear power. Chernobyl had become the rope that tied the industry together across political and cultural boundaries.[2]

Lukonin and Marshall began the ceremony, each signing the Charter and then shaking hands for the photographer, a symbol of the cooperation, collaboration and dedication that brought the nuclear industry together and that those present hoped would be a major part of its future. Then, one by one, each delegate came up to the stage, signed the Charter, shook hands with Lukonin and Marshall, posed for the obligatory photo and returned to his seat. For nearly two-and-a-half hours, some 140 delegates from 29 countries representing more than 146 nuclear operators marched to the stage, pledging their support for the new organisation. Throughout the speeches, simultaneously translated into six languages, few left the auditorium. When the last delegate signed the Charter, the room burst into applause – from a sense of considerable accomplishment and great relief.[3]

"It was a wonderful moment," recalled one organiser instrumental in the planning of the meeting. William S (Bill) Lee III, an industry veteran of 38 years, the head of Duke Power Company, and soon to be President-elect of WANO, noted as he watched the delegates sign the Charter, "I was overwhelmed with the most remarkable feelings of being involved in a major evolution of the course of history. Here I was, present in the room where people from all over the world who were responsible for the operation of 420 power reactors were pledging, with deep sincerity and emotion, to work together toward the safe operation of nuclear reactors and the benefit of mankind." For Lee, the signing ceremony represented not only what "WANO could mean to all of us, and not only because of its promise in the nuclear field, but as a model for cooperation in other areas". WANO's birth was greeted with great enthusiasm and with substantial expectations. Matching the promise would be no easy task.[4]

****

The single event that had assembled the world of nuclear power in the spirit of cooperation was the catastrophic accident at the Chernobyl nuclear power plant in Ukraine on 26 April 1986. An explosion and fire at the Number 4 unit released massive quantities of radiation that fell over large parts of the western Soviet Union and Europe, resulting in the evacuation of the nearby city of Pripyat and an international outcry regarding the safety of the Soviet nuclear power industry.

The accident shook the foundations of the Soviet government as well as its nuclear power programme. Although a scientific team departed for Chernobyl the day of the accident, the Soviet government, with a tradition of operating in secrecy, made no public announcement until a Swedish monitoring station detected the radioactive plume two days later. The initial lack of communication after the event was claimed to be due to a lack of awareness about the severity of what had happened. But once announced, the explosion and its aftermath "made absolutely clear how important it was to continue the policy of *glasnost*", Gorbachev later explained. Chernobyl, in his opinion, proved the wisdom and necessity of the openness policy and allowed change to occur. After breaking the silence surrounding Chernobyl, one observer noted, Soviet announcements were marked thereafter by "honesty and unparalleled information".[5]

If Chernobyl had a notable effect in terms of how the policy of glasnost was pursued, it also put nuclear power operators throughout the world on notice that a tragic accident to one plant had serious repercussions for them all. "It has taught us a lesson of stark self-interest," Lord Marshall warned, "that a nuclear accident has an economic fallout which spreads even wider than the radioactivity which is released." Although the Americans had learned this lesson in 1979 as a result of Three Mile Island, neither that accident nor the US utility industry's response had resonated much outside North America. As a result, before 1986 there was little sense of collective responsibility among nuclear operators in Europe and Asia. Most considered their operations to be well run and safe. "There was a lot of complacency in Europe; we know how to do it here," one European nuclear power official recalled about that period. "The general feeling was that the standards in America weren't as good as they should have been." Among nuclear plant operators in Europe and Asia, Three Mile Island was "something that happened over there and not something that will happen to us".[6]

In addition to an attitude of superiority, the culture of nuclear plant managers outside the US operated against any core belief in collective responsibility. Any lessons regarding the safe operation of reactors were rarely shared within a company, and never beyond. "The idea that a power station manager might actually tell somebody that he had made a mistake so they could learn from it was completely anathema," according to Andrew Clarke, one of WANO's earliest employees. "There was great reluctance to get any exchange of experience among operators at the time. The view was that you couldn't share experience between different types of reactors, so there was very little exchange between the UK operators and the rest of Europe for that incorrectly perceived reason." Consequently, there were no mechanisms for capturing and sharing operational experience. Chernobyl, however, served as the catalyst to dissolve the arrogance and isolation of the old order and begin the process of preserving a nuclear future.[7]

The road from Chernobyl to the Inaugural Meeting in Moscow ran through the US and France. The initial impetus for some type of worldwide response came from the Institute of Nuclear Power Operations (INPO) in Atlanta, Georgia. Established in 1979 by the US utility industry in the aftermath of Three Mile Island, INPO was a non-governmental organisation that sought to promote the highest levels of safety and reliability in the operation of commercial nuclear power plants. In its formative stages, INPO found it difficult to overcome the industry's reluctance to share operating information, especially if it exposed any weakness, and a general opposition to add another layer of operational safety and performance review beyond the US Nuclear Regulatory Commission. But as the organisation took shape, it demonstrated the value of peer reviews, operator training and the exchange of operating experience in preventing accidents and improving performance. Most US utilities came to see INPO as a good investment. By 1986, the concept of self-regulation had proved effective and gained broad support from the industry and from the Nuclear Regulatory

Commission. INPO officials, led by Chairman and CEO Dr Zack T Pate and President William S Lee III, immediately recognised the historical parallels between Three Mile Island and Chernobyl and believed that the INPO model, or some version of it, would be applicable on a global basis.[8]

Shortly after its formation, INPO had explored bringing foreign utilities under the INPO umbrella, hoping to tap their operational knowledge and expertise through an international participant programme. The first company to join was Électricité de France (EDF) in April 1981. By the end of 1983, INPO's International Participant Program (IPP) included nuclear utilities from 15 countries in North America, Europe and Asia, comprising half of the world's nuclear operators outside of the Soviet Bloc. In May 1986, INPO's Board of Directors decided to create two groups – one from the Board of Directors, the other from the IPP – to investigate how best to respond to Chernobyl. Pate asked Bill Lee, who had been instrumental in the creation of INPO, to chair the Board sub-committee to consider an appropriate response to Chernobyl. Lee believed that the solution was to open up the IPP to all-comers and his committee pushed for some way to accommodate additional foreign nuclear operators. But Pate was uncertain whether European or Asian operators would join a US-led initiative. Pate asked Stanley J Anderson, a retired Admiral who headed INPO's international programme, to work with the new IPP Chair, Andrew Clarke of Great Britain's CEGB, and the previous IPP Chair, Thomas Eckered, a Swede who had just joined INPO with the assignment to open an INPO office in London, to form an International Participant Advisory Committee (IPAC) and examine other options.[9]

Clarke and Eckered recruited IPP members Juan Eibenschutz, a Director of the Mexican Comisión Federal de Electricidad (CFE) and Vice Chair of IPAC; Boris Saitcevsky, who was EDF's representative to the Union Internationale des Producteurs et Distributeurs d'Énergie Électrique (UNIPEDE), an alliance of European utilities; and H Chu of

Taiwan Power to provide a wide range of opinions from a worldwide perspective. Eibenschutz, who was well known in the international community, came from a small nuclear programme. His role was to emphasise that any new organisation "was not just for the big boys", but would be of value to small nuclear programmes throughout the world and that they should support the initiative. Saitcevsky, a small, wiry man who could offend others on the Planning Committee by pushing his own personal views and ambitions, was needed for his connections in the Soviet Union. Chu, as it turned out, was ill and sadly died before the committee presented its recommendation.[10]

Eckered's involvement indicated the depth of INPO's commitment to nurturing an international response, whatever path it took. Clarke's group opposed Lee's expansion of the IPP, arguing that any nuclear operator that had not already joined a US organisation probably wouldn't do so "because for whatever reason or other, they didn't like working with Americans or being dominated by them." In short, there could not be an expanded INPO or an international INPO. It could not be a US organisation. There had to be a more international approach. Lee and the Board sub-committee agreed and backed off. The result of these discussions was to convene an International Nuclear Utility Executives Meeting (INUEM) somewhere other than in the US to "address methods of enhancing cooperation between nuclear operating organisations worldwide".[11]

The decision in the US closely paralleled concurrent discussions in Paris. EDF, the world's largest nuclear utility, had been shaken by the Chernobyl accident and had discussed options for a response without settling on a single course of action. Pierre Tanguy, EDF's representative to IPAC, reported on the discussions in Atlanta and the possibility of sponsoring a meeting in Paris. The executives, including EDF Chairman Pierre Delaporte, Deputy Manager Rémy Carle, and Jacques Leclercq, the Executive Vice President for Generation and Transmission, embraced the idea and soon adopted it as their own. In the fall of 1986 Leclercq met with Pate and mentioned the idea of EDF hosting an organisational meeting in Paris. Pate could not have been more pleased. Between the two parties a general plan of action was hammered out to hold a meeting of the world's nuclear executives in Paris the following year.[12]

Aside from being one of the most attractive cities on earth, Paris had several advantages. Air service was frequent and convenient from most parts of the world. A more neutral site than nearly anywhere else in Europe, the city would be acceptable to the Soviets, whose participation everyone believed was essential to the success of a future organisation. And EDF would absorb the costs of the meeting. That offer was crucial as it allowed organisers to invite people from around the globe without having to pay more than travel expenses, making it far more likely that invitees would attend. INPO would assist with staff and other resources to organise the meeting, but otherwise remain in the background. The planning group seized on the agreement. For an industry shattered by Chernobyl, nuclear operators could establish guidelines for restoring confidence in commercial nuclear power and reinvigorate the industry by creating a new era of self-strengthening cooperation. To rise from the ashes of disaster, nuclear power producers would draw on the lessons of Three Mile Island and Chernobyl to usher in the reforms necessary to make nuclear power safer and more acceptable worldwide. Paris, the City of Light, would host those who produced the power to light many of the cities throughout the world.[13]

The organisers, however, faced a major problem planning a meeting of the world's top nuclear executives. First of all, they assumed that utility executives would be too busy to spend much time at a meeting, even in Paris, so they anticipated just two days of sessions to get the organisation off the ground. That meant that most of the decisions had to achieve a consensus on a broad plan of action to move forward with a new organisation before the delegates gathered. However, for an industry that rarely fostered communication or collaboration, no global listing of nuclear operators existed. Although utility executives could be readily identified in North America, Western Europe and those countries that had joined IPAC, there were large gaps when it came to identifying individuals in Latin America, Eastern Europe and parts of Asia. Not only did the organisers not know which entity ran some of the world's nuclear power stations, they did not know the names of the top utility executives they wanted to attend the Paris meeting. According to a staff member at the time, "we didn't know who to write to for over half of the operators of nuclear power plants in the world." And so began the difficult work to determine who, exactly, would be invited.[14]

The staffing for a Planning Committee and funding to organise the Paris meeting came from INPO, which had the experience, commitment and financial muscle to make it happen. Pate assigned Anderson, the Director of International Operations for INPO, including IPAC, to start the process. Although a retired admiral, Anderson was a slight, quiet, unassuming man. "If you saw him in a crowd," one colleague said, "you could easily pass him by." But appearances were deceiving; his manner belied his effectiveness. "He never shouted, never needed to shout. If Stan said something, people listened. His word was authority." Anderson was very important to the workings of the Planning Committee. He hired George Hutcherson, a US Naval Academy graduate and former submarine officer, to add technical expertise to his staff. A native of Richmond, Virginia, Hutcherson had worked at government nuclear facilities in the western part of the US after leaving the navy and saw the INPO position as a way to move back east. By the end of 1986, Hutcherson had moved to Atlanta. His first assignment was to determine the names of the operators to invite from Eastern Europe.[15]

Hutcherson would soon join another Planning Committee member, Thomas Eckered, who in January 1987 had moved to London to open an INPO office. The new office provided additional support for the upcoming organisational meeting in Paris. For the first few months the office was on the ground floor of his home. Because of a communication workers' strike, he had no telephone for several months and resorted to making calls in the evening from several of the ubiquitous red phone booths in the neighbourhood. He solved the problem by purchasing one of the first mobile phones. Several months later, INPO moved into Chelsea Chambers at 262a Fulham Road in Kensington. A former church administration building, the structure had been redeveloped by a Swedish company into small offices. INPO occupied two office units, sharing a fax machine and a secretary with nearby occupants. The leased space also had phone lines and, soon thereafter, access to the public data network and an INPO nameplate.[16]

Thomas Eckered was trained as an aeronautical engineer and worked for Saab, the Swedish aircraft manufacturer. He spoke several languages and left Saab for the Swedish Foreign Service at the Organisation for Economic Co-operation and Development (OECD) in Paris, where he worked with a number of European utilities – an appropriate background that included the linguistic and diplomatic skills INPO believed necessary for the tasks ahead. In addition, Eckered was committed to nuclear safety. After Three Mile Island, Swedish nuclear utilities had organised themselves for cooperation, establishing an INPO-like organisation called RKS to be a clearing house for reactor operating experience in the Scandinavian nation. As the Director of RKS, Eckered had joined the IPAC. However, in late 1986, RKS merged with another organisation and Eckered found himself out of a job. Pate seized the opportunity to hire Eckered and open an INPO office in Europe. Eckered was the only non-American employed by INPO and believed that his London posting was always intended to be the international coordinating centre. By the summer of 1987, Hutcherson, who had been working with the Planning Committee from Atlanta, was on his way to London to join the team.[17]

The third individual critical to the Planning Committee was Andrew Clarke, who had replaced Eckered as chair of the IPAC. Clarke had been instrumental in shaping the direction the INPO working group had taken toward a new international organisation rather than an expanded INPO. Born in London just as the war ended in Europe in 1945, Clarke took his degree in applied mathematics at Cambridge University and accepted a position with the CEGB, which had been created in 1957 to supply energy in England and Wales. Clarke possessed a quick mind, an articulate tongue and a talent for organisation. He was a keen observer of people and read others well. In January 1987, Clarke chaired the first Planning Committee meeting in London. He was an ideal choice – he ran meetings well and kept the discussions on point, a highly valued skill for any group, but especially one in its infancy. His boss, CEGB Managing Director Lord Walter Marshall, supported the idea of an international organisation and allowed Clarke to devote some of his time to get it started.[18]

Over the next few months, all three men, Hutcherson, Eckered and Clarke, were in London working on plans for the Paris meeting. With assistance from INPO, UNIPEDE, the International Atomic Energy Agency (IAEA) and British embassies throughout the world, the small staff was able to assemble an invitation list of the top executives of nuclear operators in 31 countries. The key to creating a viable organisation was to convince all the world's nuclear power operators to participate. This meant putting aside national rivalries and mistrust to coalesce in a common good. Many European nuclear operators resented INPO, in part because it was American and in part because they believed that American standards were not particularly high and that nuclear plants in Western Europe operated more efficiently and more safely than US plants. In addition to distancing themselves from the Americans, "there was a fair bit of rivalry between the other utilities in Europe and EDF, which they saw as a bit of a bully," recalled one early staffer. "If EDF says it's right," he recalled, "the other utilities thought "it must be wrong. We had to overcome that."[19]

Aside from dealing with the jealousies and animosities among the Western operators, the planners, to build a viable organisation, had to ensure the participation of the USSR and Japanese nuclear plant executives. Because of Chernobyl, Soviet participation was obviously crucial to the success of any international nuclear safety federation. Yet after the accident, the rest of the world was uneasy about the differences between the RBMK and Western nuclear units and the European response was more critical than constructive. Implementation of glasnost was uncertain. "So it wasn't obvious that the idea that the Russians would then come and join an organisation set up by the West would work," Clarke recalled. The same was true for the Japanese. Prior to Chernobyl, Japanese nuclear power companies did not participate in any kind of international organisation. After the formation of Japan's Central Research Institute of the Electric Power Industry (CRIEPI), modelled on the Electric Power Research Institute in the US, the Japanese joined the IPP, assigned an engineer to work at INPO, and sent a representative to the IPAC, but remained aloof from any deeper international involvement. [20]

INPO, with its tradition that nuclear operators were part of a chain only as strong as its weakest link, quickly sought a dialogue with the Soviet Union. Soon after Chernobyl, INPO officials began discussing with Soviet officials the possibility of joining INPO or an INPO-like organisation to improve nuclear reactor safety. In November 1986 Bill Lee and Stan Anderson travelled to Rome to meet with Professor Evgeny Velikhov, a noted Soviet physicist who would later head the Kurchatov Institute, but nothing grew out of that contact. To convince the Soviets, the London planning group decided to sidestep direct American involvement and to approach the Soviet officials through two European organisations with which they were familiar, the IAEA and UNIPEDE. Although the IAEA's main focus was on nuclear non-proliferation, another, though lesser, function was inspecting nuclear power plants throughout the world through its Operational Safety Review Team programme, or OSART; at a country's request,

OSART sent a team of experts to evaluate the operational safety of a nuclear power plant. Over the years, the IAEA averaged about two OSART visits a year. The Soviets had worked with the IAEA and were comfortable with its degree of supervision. Therefore, it was important for the IAEA to support the proposed initiative. One stumbling block was the IAEA's senior staff, who viewed the idea of a utility-led organisation as a threat and opposed it. To gain credibility with the Soviets and assure the IAEA that an industry group would not endanger the IAEA's operations, Eckered and Clarke went to Vienna to speak with Dr Hans Blix, the director general of the IAEA. Eckered knew Blix when he was the Swedish minister of foreign affairs and when he and his party had supported the retention of nuclear power in Sweden after the Three Mile Island accident. In addition, Blix was the first Westerner to have been invited to Chernobyl and view the extent of the damage there. Unlike his staff, Blix immediately saw the benefits of a new international organisation. He told Eckered and Clarke that he did not have the resources to expand the OSART programme and was unlikely to get additional funding for it in the future. "If the utilities can do it themselves and we can have some access to the results," he said, "I'll be able to say to the international community that we're supporting this utility activity." On that basis Blix and the IAEA would become important allies, and their approval eased Soviet concerns about participation.[21]

UNIPEDE also offered an opening to the Russians. Based in Paris, UNIPEDE, primarily an information-sharing operation, drew its membership from the electricity supply industry throughout Europe and Asia, and had recently developed an interest in nuclear power producers, in part due to its relationship with EDF. That company's representative to UNIPEDE and Director of its Management Committee was Boris Saitcevsky, whose family had emigrated from the Soviet Union to France years before. Saitcevsky spoke fluent Russian and had built strong links to the Soviets through UNIPEDE; he was soon commuting to London to assist with the

Soviet connection.[22]

To bring Japanese utility executives to the meeting, the organisers believed they had identified the one individual who could do it – Lord Walter Marshall of Goring, the head of the CEGB and a man of international stature in the field of nuclear energy. Marshall was a highly regarded and widely published nuclear scientist who had served as Chairman of the UK AEA. In that capacity he had befriended many in the Japanese nuclear community. His opinion carried great weight in Japan. The Japanese greatly respected Marshall's intellect, technical expertise and scientific accomplishments and treated him accordingly. He once told his aide, not entirely in jest, that "they treat me in Japan like I think I ought to be treated at home." When Clarke approached him about becoming involved in a new international organisation, however, Marshall declined. He could support the concept, but understood the difficulties inherent in creating and operating an international organisation. He would not commit personally to an organisation he thought might fail. At the urging of INPO's Pate and Lee, who realised that no American could lead an international industry group, Clarke returned to Marshall to ask him to chair the Paris meeting. Again, Marshall said he did not wish to be associated with something that was doomed to fail. "That's a bit disappointing," Clarke recalled telling him. "What will I tell all those people from America, from France, from Japan who have said you're the only person in the world that could possibly make this succeed? I'll have to go back and tell them you won't do it." The appeal went directly to Marshall's ego. "I suppose that if they all think I'm the only one to do it, then I'll have to do it, won't I?" With Marshall's involvement, the Japanese were likely to follow. With the meeting chaired by a man of Marshall's prestige and international stature, the organisers hoped his name might attract other top-level attendees as well. The first organisational steps had evolved into strides on the road to Moscow.[23]

At the end of March 1987, invitations to the Paris conference went out to nuclear utility executives in more than 30 countries. Using the IAEA and UNIPEDE connections to woo a Soviet contingent to the Paris meeting, the invitation letters were signed by Clarke and Saitcevsky, representing IPAC and UNIPEDE/EDF, respectively. Clarke was pleased with the letter, which was the culmination of the committee's months of preparation – Marshall would chair, Blix would speak and EDF would pay. Letters went to every nuclear operator and nearly all responded. In June, Marshall sent a second letter to the invitees outlining some ideas describing how the senior executives might work together and asking them for any ideas that they wished to contribute. The response was, in Marshall's view, "very positive, a clear sign that the time is right for this meeting and a clear sign that there is a general wish to set up a new organisation." He was correct on both counts.[24]

Nevertheless, there were a couple of bumps along the road. Ian McRae, the head of South Africa's nuclear company, Eskom, could not travel on a South African passport because of other nations' response to that country's apartheid policy. McRae worked out an agreement whereby he would travel to the meeting on his British passport, but not speak during the conference. The second issue concerned China and Taiwan. As a condition for attending, the Chinese insisted that Taiwan be listed as Taiwan, China – not simply Taiwan. The French had wanted attendees to have national flags in front of their seats, an idea that also caused an issue with both China and Taiwan. The organisers solved that dilemma by saying that the meeting was of utility executives, not countries, so Taiwan Power and the Chinese National Nuclear Corporation agreed to attend as utilities, not national entities. The idea for flags around the table quietly died. The most serious bump of all, however, turned out to be the Soviets, who did not respond to the invitation. East Germany, Czechoslovakia and Bulgaria had notified Clarke's committee that they would not be attending, but the Soviets were silent. About three weeks before the executives were to gather in Paris, Saitcevsky had a phone call saying that a delegation headed by Nikolai Lukonin, the Soviet minister of nuclear power, would be attending and would be pleased to speak. And the East Germans, Bulgarians and Czechs would be coming as well. With a great sigh of relief, the committee rushed to list the new speaker on the programme.[25]

****

The utility executives gathered at the opening session on Monday morning, 5 October 1987, at the Maison de la Chimie, or House of Chemistry, a grand building on the rue Saint-Dominique with a large auditorium, ample meeting rooms and pleasant gardens for informal discussions during breaks in the sessions. Lord Marshall was thrilled with the attendance. Thirty nations were represented, sending more than 130 delegates, including 26 from the US, 15 of whom were either the chairman or president of their company. One American, the chairman and CEO of Southern California Edison, failed to attend, Marshall explained, due to an earthquake in Los Angeles. "Tell him I do not think that was a good enough excuse," he joked, setting a collegial tone for the meeting. A polished speaker, Marshall relished his role as chair and he quickly took over, guiding the agenda while assuring attendees of the important part they had already played through their letters in helping him shape the form a new organisation might take.[26]

At his best, Marshall could be all things to all people and he was at his best presiding over the conference, helped immensely by his sense of humour, which was skillfully self-deprecating and ego-boosting at the same time. "You know they asked me to be the chairman of this meeting and I have two great qualifications that make me the right person," he told the delegates. "The first one is I am not French, and the second one is that I'm not American." Joking aside, Marshall was exactly right. His duty as chair was to bring the delegates into a larger proposition, a world organisation for nuclear safety. To succeed, he counseled, the executives should put aside politics and focus on the requirements of utilities. "We need not spend much time discussing whether we should improve co-operation, but concentrate our efforts on discussing how we should go about it." That can be done, Marshall promised his audience, "without unnecessary duplication of existing efforts and without bureaucracy", reassuring friends of the IAEA and those with limited funding. "The prime responsibility for nuclear safety lies with us," Marshall reminded the executives. "This meeting provides a unique opportunity. I hope that we will all leave here tomorrow confident that we have taken full advantage of it."[27]

Indeed, much of the basic work had been accomplished by the Planning Committee in the months before the meeting convened. The committee had drawn many of its basic ideas from the INPO experience because of its extensive operational history as an industry-led safety organisation. Pate and Lee pressed the importance of top-level executive involvement if the organisation were to succeed. INPO also stressed the value of information exchanges and its Nuclear Network® system, a programme similar to that established by UNIPEDE on a smaller scale in Western Europe, and one that provided for technical exchange and special assistance visits, and analyses of nuclear plant event reports to identify the potential for future accidents. INPO also pushed publications providing performance objectives and criteria, best practices, operator training and accreditation, and, most important to the INPO leaders, periodic peer reviews. Nevertheless, whatever ideas might be adapted from the INPO experience, none of the planners sought to impose the INPO programmes on a worldwide industry group. Any full agreement to move forward would have to emerge from a consensus of all parties, and the pre-meeting communications had laid some of that groundwork. The central purpose of the meeting of senior utility executives as outlined in the letter of invitation was "to define a mechanism for the association, cooperation and commitment among nuclear utilities worldwide and to explore ways to put such a mechanism together." By the time the utility executives met, nearly all agreed that a new industry organisation was needed and had settled the basic outline of its structure. In his opening speech Marshall reinforced the suggestions from the Planning Committee that the majority of delegates had largely accepted – a common mission to maximise the safety and reliability of the operation of nuclear power plants, the establishment of four regional centres and a small central coordinating office, and the appointment of a Steering Committee to define the specific tasks to be undertaken. One other decision was left to the meeting – choosing a name for the organisation.[28]

Perhaps the most crucial innovation to help unify a worldwide organisation of nuclear utilities was the creation of four regional centres and a central coordinating office. Early in the planning stages, it became apparent that a federal, rather than a centralised, structure was the only way to gain wide support. The strong, centralised INPO structure had few adherents outside the US. To unite a fragmented industry – separated by language, reactor type, political differences and geographic distance – the concept of regional centres provided some degree of local control and assurance in an industry that after Chernobyl saw itself under attack from hostile governments and activists who threatened its very existence. While nuclear operators worldwide accepted the idea of nuclear safety, each regional culture – Western European, American, Eastern Bloc and Japanese – had its own idea of how that should be achieved and, subsequently, often rejected the approach to safety advocated by the other cultures. The Planning Committee chose carefully. The original idea for the four centres was the brainchild of Anderson and Clarke, who saw "that we needed an organisation where there was a measure of individual control and choice, and we wanted particular groups to feel ownership. We also needed some coordination, but not central direction, which a lot of places in the world weren't ready for at the time," recalled Clarke. Without granting a large degree of local autonomy, the organisation would have lost broad support and collapsed on the drawing board.[29]

The scheme recognised the four geographic and cultural regions of nuclear power, creating a loose alliance, giving autonomy and control to each within the broader confederation. To best draw on INPO's extensive experience of operating an industry safety group, one centre would be located in Atlanta. The Planning Committee expected North American utilities to naturally gravitate toward Atlanta Centre, but utilities were free to join any centre or, if they so chose, more than one. There would be a centre in Paris that would represent the nuclear operators in Western Europe. Moscow was proposed as a third centre to serve as the umbrella for all Soviet-built reactors in Eastern Europe and the Soviet Union. In Asia, a centre would be opened in Tokyo, an acknowledgement that the Japanese were veterans at operating nuclear power plants and their practices might serve as a model for those nations in Asia and the sub-continent who were beginning to build such plants. There would also be a coordinating centre, but it lacked any management or oversight authority and was designed to serve largely as a clearing house for information. To ensure broad acceptance, the regional centres would have both organisational autonomy and authority. Though connected by the common thread of nuclear safety, the regions did not have a common solution. This structure ensured that all the parties would have a more or less equal voice in Paris and a measure of regional control in the future.[30]

Despite the promising start, Marshall was correct in his assessment that a new global organisation might fail. Historically, international industrial organisations had formed as cartels, such as in the diamond and oil industries, that promoted pricing arrangements, not operational safety. There were few precedents upon which the Planning Committee could draw in shaping the organisation. All agreed that, whatever lessons could be applied from the American experience after Three Mile

Island, the centralised INPO organisational model would not be appropriate. The committee could broadly outline a structure, but the final shape would have to emerge from the delegates' comments in Paris and the continuing discussions that would follow. Moreover, there was no guarantee that an industry that lacked any history of collaboration would come to an agreement or, if it did, successfully implement it. The letters the utility executives wrote to Marshall in advance of the meeting expressing their concerns and offering their suggestions were decisive – indicating for the first time that the nuclear industry could communicate on a worldwide basis in, what Marshall termed, "a spirit of cooperation".[31]

Marshall skillfully led the executives through the broad framework of what he hoped the meeting would accomplish, suggesting, a bit artfully, that their letters had become a large part of his thinking. "Perhaps my proposals are too ambitious," he began, but "in making these suggestions I am reflecting the views you have given me." Everyone agreed that industry collaboration was needed, yet without creating a bureaucracy "which would duplicate existing arrangements". He offered a definition of the new organisation that all could also agree on: "To maximise safety of the operation of nuclear power stations by exchanging information, encouraging comparison and stimulating emulation between the utilities operating nuclear power plants." It was a definition that Marshall believed provided an umbrella broad enough to offer something for all. He applauded the idea of four regional centres and a small central centre to coordinate "the activities of the main centres", thereby alleviating the fears that a central office might control the organisation, a model feared by those centres not located in Atlanta. He called on the executives to establish a Steering Committee before leaving Paris to implement the creation of the organisation and the tasks that it should undertake. Finally, he asked them to settle on a name for the new organisation "that instantly conveys the picture of nuclear utilities working together".[32]

By the end of the first day of extended discussions, two themes emerged: cooperation and a willingness to exchange information among the worldwide nuclear community. Whatever the particular differences, the utility executives agreed that there were common goals, and "we have a joint responsibility and obligation as a nuclear community" to achieve them. There was near unanimity that the utilities should unite in the name of safety. Nevertheless, organisers were aware that the participation of the Soviet Union was of paramount importance to the success of the nascent organisation.Throughout the first day the Soviets had been supportive, but non-committal about the future. With the need to reach a decision to move forward the next day, Marshall's allies at EDF and INPO urged him to convince the head of the Soviet delegation, Nikolai Lukonin, of the critical importance of his country's participation.[33]

Marshall's courtship of Minister Lukonin began that evening. To show Paris in its best light, EDF sponsored a dinner cruise on the Seine for the delegates. Marshall arranged to sit with Lukonin. During a spectacular meal, with splendid wines and glorious views of the city as only the French and Paris could offer, Marshall pressed hard, explaining how important the Russians would be and how necessary it was for them to be part of the new organisation. Marshall was at his persuasive best. He spoke to the great need for a regional centre in Moscow, suggesting that it was an ideal solution for increasing the safety of reactors in Eastern Europe under Soviet direction. With Soviet participation so crucial, Marshall asked Lukonin to host the inaugural meeting in Moscow the following year. Lukonin smiled and nodded, but one could not be certain of the translation and the degree of understanding. In any case, the Soviet minister did not reach his high position by playing his hand early. Marshall realised that Lukonin would need to check with Moscow before making such important decisions. At the end of the cruise they parted with a cordial goodnight. Marshall had performed superbly; now the future lay in Lukonin's hands.[34]

The next day, in unambiguous terms, Lukonin demonstrated the effectiveness of Marshall's attentions. Addressing the entire conference, Lukonin explained the Soviets "must do all we can to improve our safety record. We consequently support the formation of a Union for the Safe Exploitation of Nuclear Power," whose main role was a forum for "exchanging experiences of nuclear power station operation. We realise only too well that the safe and reliable operation of nuclear power stations throughout the world will enable us to restore public faith in nuclear power." He wished to be certain that any new organisation would not duplicate the activities of the IAEA, but "work in close contact" with it. The Soviet Union, he said, supported the regional centres for the organisational structure and favoured a small coordinating centre located in Vienna. Soviet insistence on Vienna had become something of a sticking point during the conference, and Marshall's staff wondered if it would be a deal-breaker. Lukonin dismissed that concern. "We have no objection," he continued, "to its location being decided by experts nominated by the various countries concerned." It was, as Marshall said, "a historic statement." Lukonin had assured the delegates that the Soviet Union would support the new union, all but guaranteeing its creation.[35]

With the backing of the Soviet government, Marshall could express his confidence in the future. Blix had offered support from the IAEA, Pate had pledged the full support of INPO from the beginning, and EDF and UNIPEDE contributed significantly as well. Only Japan had shied from making a formal commitment to the organisation or to setting up a Japanese regional centre. The leader of the Japanese delegation, Kenichiro Matsutani, though supportive of the idea, would not act until assured of the involvement of the USSR. Just before the conference ended, Matsutani accepted the idea of "establishing an Asian Centre in Japan. For that purpose, we expect the active cooperation and participation from the neighbouring countries. That is all." Although it was unclear to some just what the Japanese were accepting, Marshall expressed his pleasure with that limited commitment. Nevertheless, the vagueness of the Japanese response presaged future misunderstandings.[36]

The historic meeting ended before lunch on the second day. Unquestionably, Marshall had led the show and though he had not obtained the full agreement he hoped to achieve, much had been accomplished. Most importantly the delegates, including those from the Soviet Bloc nations, welcomed the broad idea of an international industry group to improve nuclear safety. Following the conference, it was Lukonin who underscored the delegates' unanimity to the waiting reporters who had been invited to spread the news of the proceedings. "We must use every available means to improve and maximise the operating level of our plants; we absolutely must achieve the highest level of reliability and safety. All the participants are unanimous about this point." That show of unity was not insignificant. In addition, the attendees had agreed to establish four regional centres and a coordinating centre toward achieving the goal of nuclear safety. Pate, who had been partially responsible for INPO's successes, understood and appreciated Marshall's leadership. "We are all in debt to Lord Marshall," Pate told the delegates, "not only for chairing this meeting, but for the extensive effort he has put into planning for the meeting and into conceptualising how we proceed after the meeting. We are all beneficiaries of his interest and great wisdom." Further, he called upon the delegates to encourage Marshall to "serve as the senior statesman, guiding the initiatives which follow this meeting." Marshall, who had initially opposed his involvement, now was central to its success. He jokingly told Pate that "I particularly liked the remarks where you said what a super chap I was." Marshall had become an integral part of the fabric of the new association.[37]

Nonetheless, after a day and a half of deliberation, the name for the new organisation, like the site of the coordinating centre, remained elusive. Delegates shied from using the term organisation. "We are particularly nervous about the use of the word 'organisation'," Marshall explained. "Although it is an organisation, it portrays the wrong image and I think we should not do that for fear that we just get misunderstanding." Other speakers tried out a variety of terms. Blix suggested a "Utility Association" to differentiate it from the government-oriented IAEA. Leclercq called it the "World Nuclear Utilities Federation," though he also used the term "association". Lukonin suggested naming the new organisation either the "Union for Safe Nuclear Power" or the "Nuclear Power Station Operating Institute." The Bulgarian delegate offered the "Council for the Organisation of the Work of Regional Centres," emphasising the major operational role the centres were to hold. That suggestion caught Marshall off guard. He had thought more along the lines of a "name the instantly conveys the picture of nuclear utilities working together." Although the Bulgarian's suggestion garnered no support among the utility executives, it did indicate the division between those who supported sovereign regional centres and those who believed that some measure of centralisation was necessary to achieve the idea of working together. The compromise to establish four autonomous centres and a very subordinate coordinating centre was decisive in reaching a consensus among the nuclear executives and establishing the organisation. Eventually, differences among the centres would expose the weakness of confederation, but that disruption lay in the future. Whatever the level of agreement in October 1987, nearly everyone realised that there was considerable work to be done in hammering out the details of programmes and governance prior to the Inaugural Meeting scheduled for 1989.[38]

Marshall, INPO, EDF, and the Planning Committee staff had accomplished much. They had brought together more than 125 nuclear executives from all but two of the countries that operated nuclear power plants – Romania and Pakistan were the missing pair. The organisers had created a worldwide industry group with few prior connections and considerable political animosities stemming from World War II and the Cold War. In addition, they had gained commitments from all the nuclear power operators, as well as the IAEA, for an industry association; completed a draft Resolution establishing the principles and mission of the future organisation; and secured an agreement to hold an inaugural meeting to formalise the organisation the following year.

In closing the meeting, Marshall outlined the next steps he believed needed to be taken. With so many details yet to be determined, Marshall's summary was a model of vagueness, but also a call to action. "What we are going to do is try to produce a harmonisation between [what EDF and UNIPEDE do] and what is done in INPO and what is done in Moscow and Tokyo." Marshall emphasised what all did agree upon: the coordinating centre would be small, staffed by no more than four people. As for the location of the coordinating centre site – London or Vienna – he pushed that decision off to a future time. "We should set that matter aside until we get to know one another a little bit better." In the meantime, with the press of time to get the new group organised, Marshall intended to use the Steering Committee, an expanded version of the Planning Committee, and the small INPO office in London to get the work underway. He asked the Soviets and the Japanese to send additional help for that effort. A concern of the East Germans – the limitation on the exchange of information and data processing with the West, a result of a Cold War embargo on computer hardware and software to Eastern Bloc nations – Marshall could not solve, but would have a "group of experts" investigate. Marshall, a pragmatist, understood the problem. "I think it we would look a little bit foolish if we say we want to collaborate internationally and then we discover that we have all made such choices of computers and data format that, in practice, it is difficult to do so."[39]

While he pushed many organisational details into the future, Marshall had convinced nearly all the delegates that they had played a critical role in the development and planning of a new organisation. He had listened to and considered the views of all delegates. Few had ever enjoyed such attention on a world stage. Marshall was a consummate leader in building a broad consensus. Like a maestro, he shaped the cacophony of the delegates' views into a recognisable melody. It was not yet a symphony and, perhaps, he had inherited an orchestra that would never play one. But as a first rehearsal it boded well for the future.[40]

Another positive omen came from Minister Lukonin. At the end of the meeting he announced that Moscow would host the next conference. Marshall and others from the West were amazed that the Soviet delegation could get permission from its superiors so quickly, but were thrilled with the decision to meet in Moscow. Nearly a quarter of a century later, Lukonin allowed that he had made the decision on his own, without telling anyone in Moscow. "I did it on my own initiative," he said. "It was important that we create such an institution." Nonetheless, Lukonin was still a Soviet bureaucrat and even a bold decision needed approval. On returning to Moscow, he sought official backing for his commitment. He went to Boris Scherbina, the Deputy Chairman of the Council of Ministers. "I explained everything and said, 'You can fire me if I am wrong.' Scherbina replied, 'No grounds for that. You were correct.' My decision was supported by everybody in the USSR government, including Nikolay Ryzhkov, the Chairman of the Council of Ministers." After Chernobyl, Lukonin explained, the nuclear power industry "was fully open for the outside world".[41]

****

For the next 18 months, from October 1987 until the spring of 1989, the Steering Committee and the Interim Secretariat planned the details of an inaugural meeting for the proposed international organisation based on the resolutions presented at the Paris meeting. Lord Marshall, who, after Paris, or perhaps because of it, had gone from being a supporter to an influential proponent, offered to chair the Steering

Committee. Its first meeting in December 1987 was held in INPO's London office, which functioned as the Interim Secretariat of the Steering Committee. The committee drew on the same staff that had planned the Paris meeting, Eckered, the office Director, and George Hutcherson, his assistant. With Andrew Clarke on loan from the CEGB, the three men undertook the task of developing an industry organisation of nuclear utility companies.[42]

The Interim Secretariat soon included representatives sent from the Paris and Atlanta Centres, Bo Sunderson of Sweden and Juan Hurtado of Spain. Eckered found the arrangement of sending engineers from the regional centres quite valuable. They had "specific knowledge about reactors in different countries and knew their own organisations and people. It helped make the whole thing work because the cultural differences between the countries involved made it quite difficult." Four subcommittees, labelled Expert Groups, assisted the Steering Committee: Expert Group One, chaired by Svante Nyman from Sweden, which dealt with communication and database issues; Expert Group Two, led by Paul Dozinel of Belgium, which was to determine the scope of WANO activities; Expert Group Three, chaired by the German Werner Hlubek of RWE, which dealt with organisational and financial matters; and Expert Group Four, headed by Boris Saitcevsky, which was responsible for planning the May 1989 Inaugural Meeting in Moscow. All the Expert Groups worked to report their recommendations to the Steering Committee by the end of the summer of 1988.[43]

The Steering Committee's staff soon resolved one loose end from the Paris meeting, the official name of the new organisation. In spite of all the discussion, none of the recommendations put forward in Paris met Marshall's dictum to adopt a "name that instantly conveys the picture of nuclear utilities working together". The original concept of an association of nuclear operators gave rise initially to naming it, not surprisingly, the Association of Nuclear Operators, or ANO. But ANO, derived from the Latin word anus, meant just that in Spanish and Italian. That acronym was quickly dropped. To avoid the reference to bodily parts, the committee, at the suggestion of Saitcevsky, added the word "World." Once the letterhead carrying the name World Association of Nuclear Operators, or WANO, was printed up and "sent out," Hutcherson recalled, "it was too hard to take back". To prevent anyone from stealing the WANO name – particularly an anti-nuclear group – the committee registered the company in the UK in late 1988, with Clarke as chairman and Eckered as secretary. Thereafter, the WANO name, which met Marshall's standard, appeared in the committee's correspondence and its letterhead and the nameplate on the door of the office on Fulham Road. There was no further discussion of the coordinating centre being in Vienna. Only London, Eckered pointed out, had direct flights from every capital in countries having nuclear reactors.[44]

Concurrent with the work of the Steering Committee, parallel efforts began in each of the regional locations with the goal of having all centres in operation before the inaugural meeting. By the spring of 1989 they had reached that goal. While there were a unifying mission and common programmes, each region approached its governance and work methods a bit differently. Paris Centre attracted members primarily from Western Europe, adopted the programmes of the Steering Committee and began operation in January 1989 with Jacques Burtheret as its first director and staff on loan from UNIPEDE. In Tokyo, CRIEPI provided office space and several Japanese utilities loaned staff. By early spring in 1988, Japan invited Asian countries with nuclear power stations to a meeting in Tokyo. Utilities from Taiwan, Korea and Pakistan attended; China and India did not. Early in 1989, the countries had developed the centre's charter and programmes and the centre began operation in March 1989. The first director of Tokyo Centre was Kinji Hoshizawa, who was assisted by a small engineering staff. In Moscow, Dr Armen Abagyan, the head of the All Union Research Institute for Nuclear Power Plant Operations (VNIIAES), took the lead in establishing the centre in his organisation's building at 25 Ferganskaya Street. He had attended the first meeting of the Steering Committee in London in December and was an outspoken advocate for nuclear safety. Dr Boris Prushinsky, who had helped assemble the first group of senior scientists to travel to Chernobyl the morning after the accident, served as the first Director. The centre began operations in April. In Atlanta, INPO provided the new WANO centre both staff and space in its building on Circle 75 Parkway. Stan Anderson, the head of INPO's international programme, became its Director. Atlanta Centre started operations in March 1989.[45]

In the late 1980s, international communication was rarely easy and always expensive. The internet was only beginning to emerge and few at the time recognised the remarkable changes it might bring. But ease of communication was essential to the basic foundation of WANO – the ready interchange of operational information. After Three Mile Island, INPO had pioneered the use of communication links to share information and accident analyses among utilities through its Nuclear Network®. Over its first years of operation, Nuclear Network® had prompted more than 125 safety recommendations to INPO members, a result that caused INPO to offer the system free of charge to WANO. Initially, the French appeared to block the idea. In a show of national pride, EDF suggested that it would be better to run the system on French-built Bull computers rather than US-made IBMs. While there was often an undercurrent of tension between the French and Americans over controlling one WANO device or another in the early years, this issue was soon resolved. No one else wished to reinvent the wheel or pay for something that could be acquired for nothing. Nuclear Network® would become the communication link among the world's nuclear utilities.[46]

The question was whether the American system was compatible with and usable across all the world's communication carriers. The Americans thought it was, but the system had never been tested in every country. In 1988, Hutcherson and Olle Nockert, a Swede working at RSU and a member of Expert Group One, set out to demonstrate the value of Nuclear Network® to prospective WANO members. In Russia, they went to VNIIAES but learned the building had no public data access line. Nockert asked about contacting the part of the Soviet government that handled international communications, but no one at VNIIAES knew about it, because as Hutcherson recalled later, "everything was bucketed in the Soviet Union at that time." Nockert, however, had an address for the state communications institute. Two hours later he and Hutcherson were in a car heading across Moscow to the communications centre. Nockert explained that he wanted to use a link between the institute and a public data network node in Helsinki. "We're sitting there and for 15 minutes there was no response at all," Hutcherson remembered. "Then a lady came in and said, 'Come with me', in English. They sit me down at a terminal and five minutes later I'm connected to the INPO computer in Atlanta demonstrating Nuclear Network®." Connecting from behind the Iron Curtain was a bit of a pleasant surprise for Hutcherson. He did a quick demonstration of what Nuclear Network® would provide and quickly left before any glitches popped up. "It was an eye-opener for us and it certainly was an eye-opener for the nuclear side in Russia." The demonstration "proved that even the Eastern Bloc could get connected to our Nuclear Network® here in Atlanta, which meant that we really did have a true way of communicating." Thereafter, Nuclear Network® was available on a trial basis in advance of the official formation of WANO in 1989. In Hutcherson's opinion, that early demonstration was vital to WANO's future. "Being able to electronically communicate was the key to WANO throughout." By proving the value of Nuclear Network® in the spring of 1988, the work of Expert Group One was largely completed.[47]

Expert Group Two took on the task of determining what type of programmes WANO should undertake. In Paris, delegates had agreed to three exchange activities:

operating experience, including event notification reports and event analysis reports; good practices; and technical assistance that included workshops, seminars, and personnel exchanges. The last came to be known as the Operator to Operator programme. The Americans had also pushed for two INPO programmes. One was the use of industry performance indicators such as industrial safety, unplanned automatic emergency shutdowns, or "scrams," collective radiation exposure, forced outage rates and capacity factors, but there was no consensus on a set of commonly accepted performance indicators, so the idea was dropped. Also eliminated was a programme of peer review plant evaluations, which met broad opposition from utilities outside the US. By April 1988 Expert Group Two presented three recommendations to the Steering Committee.[48]

To complete the process, the Steering Committee approved Expert Group Two's recommendations, and then asked its staff to write policy guidelines for each programme. Once the Committee approved the policy document, staff drafted implementation guidelines that carried the specifics for all the WANO centres. Over the next year, Pate and INPO continued to push for adding performance indicators to the WANO programme and just before the Moscow meeting convened, the Steering Committee adapted the set of indicators that had been developed by INPO. These were added as a fourth programme. Expert Group Two demonstrated the impact that INPO had on the initial programmes of WANO. The reporting requirements for WANO's operational experience programme – event notification reports and event analysis reports – were essentially the same as the INPO significant event criteria.[49]

Expert Group Three had the task of developing the WANO Charter and Articles of Association. It had the advantage of working from an earlier version of a charter that Eckered and Hutcherson had sketched out before the Paris meeting. IPAC also reviewed it. However, Pate, who also looked over the draft document, told them that it was premature and should be shelved until the Steering Committee met with the governance group in late 1987. At that point, the draft became the starting point for the development of the WANO Charter.

WANO's decentralisation posed some obstacles to the development of a unified Charter. Always part of the organisational structure, the four regional centres divided along financial, language and technical lines, and each had its own governing board. First of all, costs were regionally based. Centres collected their operating funds from affiliated utilities. Although each centre contributed a small amount toward the operation of the Coordinating Centre, most of their expenditures were in local currency, an arrangement particularly important to the ruble-based Soviet and Eastern Bloc nuclear plants. In addition, while the common language of WANO was English, the regional centres often conducted their business in Russian, French or Japanese. Reactor technology was also a determining factor in a utility's decision to affiliate with a regional centre, a factor also recognised by the Charter's developers.

A more delicate governance issue was voting. A central question was how voting rights would be determined – by the umbrella organisation such as INPO or EDF, the number of plants in a country, or by country? There was an unwritten gentleman's agreement that INPO, though it represented all the US nuclear utilities, would have one vote. EDF, with 54 plants, honoured the idea and would have only one. The Japanese formed an organisation that brought the nine Japanese utilities under one organisation to conform with the INPO and EDF approach. When the Charter was completed, voting was on a country basis. The only central governing feature was the WANO Governing Board made up of two representatives from each regional centre and a chairman, for a total of nine members. As it turned out, under the WANO Charter, the only item that all the members voted on was the president, the individual responsible for hosting the next biennial meeting.[50]

By early 1989 all was in place for the Inaugural Meeting and the official creation of WANO. The Inaugural Meeting in Moscow was the culmination of work of the Interim Secretariat, Expert Groups and Steering Committee over the previous 18 months. The regional centres were operating on an interim basis, each under its own charter and with its own staff and governing board. The result was a remarkable consensus among international nuclear power companies. The Coordinating Centre, while still suspect among centres that wished to run their own affairs, was "to ensure that there are effective exchanges across regional boundaries" and to ensure that the resource and expertise of all the members was "fully utilised to the benefit of all". Eckered had been selected as the director of the Coordinating Centre; Minister Lukonin was serving as WANO's first President; and Lord Marshall had agreed to become the first Chairman of the Governing Board. Marshall was characteristically upbeat on what had been achieved. "Every prospective WANO member," he boasted, "has been determined to find solutions rather than difficulties. We have a unanimity of purpose that is rarely seen even within the confines of a single company, let alone across 150 companies from all corners of the globe." All that was needed, Marshall said, was for "every nuclear utility in the world" to implement their commitment to work together by signing the WANO Charter.[51]

****

The signing of the Charter, central as it was to establish an industry commitment to a global federation in support of safe nuclear power, was not the only highlight of the May 1989 Inaugural Meeting. The mystery of Moscow and the Soviet Union also attracted delegates. Many of the participants had never been to Moscow. Even with the beginnings of *perestroika* and *glasnost*, obtaining a visa, travelling to and within the Soviet Union and finding suitable accommodation for the attendees was still not a simple process. In addition, the official Soviet travel agency, Intourist, still controlled access and programmes for foreign visitors. There was also some concern about surveillance, particularly from the American side. To dispel these concerns, the Soviet hosts made certain that the delegates received the very best Moscow could provide. The Sovincentr, a collection of buildings on the Moskva River – which housed hotel rooms, large halls and auditoriums, as well as reception facilities – served as the country's international trade centre. On Sunday, the day before the meeting, participants and their guests enjoyed a guided tour of Moscow and the Kremlin. That evening delegates boarded buses and, with traffic blocked to speed the motorcade along, were whisked through the city's rush hour traffic to the Bolshoi Theatre for a ballet performance. During the meeting, spouses were treated to full days of sightseeing at art museums such as the Kuskovo Museum of Ceramics, the All Russian Museum of Applied and Folk Art, and the better known Pushkin Fine Arts Museum; and were taken on trips outside Moscow to the Novodevichy Convent, Zagorsk Monastery and the Vladimir and Suzdal Museum, parts of the important historical and architectural areas known as the Golden Circle. Each evening delegates attended sumptuous receptions with tables full of traditional Russian delicacies. Lord Marshall remarked that in the UK a reception meant a few drinks, but in Moscow it meant a stand-up full meal that went on and on until the ice cream, or *moroshina*, was served. After the meeting concluded, participants could choose three-day technical tours to Leningrad for a tour of the city and the Leningrad nuclear power plant that Lukonin had once managed, or a trip to Kiev and a tour of Chernobyl and Pripyat, which could accommodate only a limited number of tourists per day. The meeting and the tours opened up attitudes toward the industry's Soviet colleagues, but also allowed participants to socialise, to meet and speak with colleagues about common concerns, be they at home or at work, on a level not seen before. The interpersonal relationships and common bonds, which INPO had found so valuable for US operators, began to form internationally at meals and during the receptions. Pate emphasised this connection, urging close association among the friends one made through WANO. "One of the key phrases in the WANO Charter," he said, is "encouraging communication, comparison, and emulation among its members." Importantly, the work of the centres, the WANO programmes and the continuation of the biennial meetings fostered this interaction and strengthened the bonds between the members.[52]

Before the delegates left for their post-meeting technical trips or to head home, they had one last business item to complete, election of the new president. Bill Lee, the American utility executive who had been critical to the formation of INPO in 1979, was elected by acclamation – a standing ovation – an indication of his worldwide reputation. Lee's acceptance speech was different from those given earlier in the meeting, which had been mostly praise for those who had worked on the conference and perfunctory recitations of a company's or country's nuclear programmes. Lee's speech was anything but perfunctory. If Marshall's role as chair was to paper over the differences among delegates in order to gain consensus, Lee's role was to test that consensus. "WANO is not a political entity," he began. "Our members profess varied philosophies and beliefs. But we have in common an absolute dependence on the safe performance of one another's nuclear plants. We must be constantly vigilant to keep our differences from impeding our safety progress. We will have to overcome a tradition of non-cooperation." He first challenged WANO's newly elected leaders, the WANO and regional governing boards. "This has been an exciting beginning, not only for what we have done here, but even more for what we expect." He urged them to build on that enthusiasm to improve safety and reliability. "Nuclear plant performance improves," he said, "when leaders set specific goals and a schedule to meet those goals." He asked each regional board to set "quantified goals for WANO regional performance to be met over the next two years" with quarterly mileposts or measuring points along the way. The WANO Governing Board should review those goals and incorporate them into a two-year system of overall goals to share with members "so we will know the specific WANO progress we can expect between now and 1991," when the Boards reported on their accomplishments during the next General Assembly of WANO in Atlanta. One particular goal he urged WANO to achieve was for each nuclear plant in the Soviet Union and Eastern Europe to be visited by a team from a Western utility, while the Eastern Bloc operators would make a return visit to a Western site by the next biennial meeting in Atlanta in 1991. Accomplishing this exchange would become a focus of Lee's presidency.[53]

Next, Lee challenged each of the delegates "to prepare your nuclear utility to share information and operating experiences" through the WANO communications network that connected each power plant to one of the regional centres and the centres to each other. "WANO's success," he warned, "will depend on how well each of us uses this network. WANO's contribution to your safety will depend on the vigour you apply to two functions—sending and receiving. If you solve a problem, others won't gain from your experience unless you share it." One key to success, Lee said, was the effective use of information. "Sharing may prevent an accident and is therefore in our self-interest. How well your people send and receive depends on you. It depends on the signal that you personally give as soon as you arrive home. You are the boss. If each of us gives this strong signal, WANO's beginning will take on reality." By being successful in sharing operating information, WANO members would build mutual trust and professional respect, Lee believed, and the strength and courage to expand WANO's programmes, such as plant evaluations and accreditation, which INPO had found so valuable, in the future. "WANO is hope born out of shame – a shame we never intend to see again. I call on everyone here to commit yourselves to the tasks in the months ahead. We must learn to trust one another, and we must work hard at it. It's up to you and me," he concluded, "to make WANO successful."[54]

Chapter Two

THE SECOND **MARSHALL PLAN**

Lord Walter Marshall of Goring: it was an impressive title for an imposing – and complex – man. While the fact that he was neither American nor French aided his selection as WANO's first Chairman, nationality was not the key criterion. Marshall had impeccable credentials. He was something of a polymath – a renowned physicist, an experienced administrator, an eloquent and forceful advocate for nuclear power and a proven, well-liked leader with both industry and international experience. He had been instrumental in the creation of the World Association of Nuclear Operators. In 1989, all agreed he was the perfect choice to lead the fledgling WANO as Chairman of its Governing Board.

Marshall's career had been meteoric. Born in Rumney, near Cardiff, Wales, in 1932, Marshall was the son of a baker and the youngest of three children. He developed a love of mathematics in primary school and a serious interest in chess, becoming the Welsh Junior Chess Champion at 15. He earned a scholarship in 1949 to study at the University of Birmingham, graduating with a First in Mathematical Physics in 1952. He received his doctorate in theoretical physics two years later at age 22 for his work on magnetism and neutron scattering under the direction of Rudolf E Peierls, a refugee physicist from Nazi Germany who had helped work out the theory for creating an atomic weapon and who joined the Manhattan Project during World War II. Upon receiving his PhD in 1954, Marshall, through his connection with Peierls, joined the

Atomic Energy Research Establishment (AERE) located at Harwell, Oxfordshire, the centre for nuclear research and development in the UK. Soon after accepting his first job, he married his Rumney childhood sweetheart, Ann Shepperd.[1]

Marshall's rise continued as if powered by destiny. From 1957 to 1959 he studied in the US, first at the University of California, Berkeley, and then at Harvard University, before returning to Harwell. In 1960 he became Division Head of Theoretical Physics and rose through the ranks to become Director of Harwell in 1968. Marshall's performance at Harwell confirmed his ability to lead organisations and motivate the work of others, leading to a switch in his career from pure science to administration. Three years later he was elected a Fellow of the Royal Society. In 1972 he joined the board of the UK Atomic Energy Authority, rising to Deputy Chairman with special responsibility for the AEA's scientific and technical policy in 1975. At age 45, he was elected a foreign member of the Royal Swedish Academy of Engineering Sciences.[2]

During the oil crises of the 1970s, Marshall became increasingly concerned with questions of energy policy and became an outspoken advocate of nuclear power, demonstrating, said one observer, "an evangelical commitment" to its expansion. He advocated building a pressurised water reactor at Sizewell and a major construction programme of nuclear power plants in Great Britain, focusing on questions of operational safety and large-scale accidents. During this period he built "an unusually warm and productive relationship with the Japanese". By 1981 he had become chairman of the AEA and was knighted the following year. A champion of nuclear power – he was known in government circles as "Mr Nuclear" – Marshall was appointed, in July 1982, Chairman of the Central Electricity Generating Board to revitalise the UK nuclear power programme. According to his biographers, "it was the start of some of the happiest years of his life." For his success in "keeping the lights on" during a protracted coal miners' strike in 1984–1985, Prime Minister Margaret Thatcher rewarded him with a life peerage. Sir Walter Marshall became Baron Lord Marshall of Goring, a town on the Thames where he lived most of his career. He was 53.[3]

Marshall's intellectual strengths and forceful personality made him a prominent participant at international conferences. *The Times* of London said that he combined "intellectual brilliance with a forthright manner and a bulky presence. His energy was formidable." His whole career, according to his close friend John Baker, "was a high-wire act, combining showmanship and skill". A self-confessed workaholic, Marshall was a man "big enough physically as well as intellectually to be worthy of caricature, and he could happily exploit this by poking fun at himself". His sharp wit and "effervescent sense of humour" were formidable weapons in his armory, helping to explain "the extraordinary affection he could command among his colleagues and staff despite his ability also to make enemies". On public platforms, Marshall deftly applied these skills, reducing the audience "to helpless laughter with exquisitely timed [jokes], deploying his larger-than-life personality with its unapologetic egocentricity…and his great sense of theatre." Baker's friend was also a man who measured his value in his salary. Marshall, remarked one former colleague, had "the aura of a man who knew his worth and could command his own terms on such matters as living and travelling arrangements." In short, he revelled in being a leader and in the perks that accompanied a high position.[4]

Marshall's first direct involvement with an international nuclear accident came in August 1986, when he led the British delegation to the special conference on Chernobyl convened by the International Atomic Energy Agency. In his analysis completed in early May, Marshall had told industry executives that the RBMK reactor was a flawed design, lacking satisfactory safety characteristics… 'He's admitting faults in the design, in training, in Soviet safety philosophy'... Marshall jumped at the opportunity to help the Soviet government. But responsibilities at the CEGB prevented Marshall from taking an active, on-site role regarding Chernobyl. Instead, he agreed to assist with the development of a proposed international nuclear safety organisation.[5]

Just before the WANO Inaugural Meeting in Moscow in 1989, a policy change in the British government to privatise the electricity industry eliminated Marshall's nuclear expansion programme. He mistakenly thought his personal connections with Prime Minister Thatcher would preserve his position, but politics prevailed and by the end of the year he found himself out of a job as Chairman of the CEGB. Officially, Marshall's departure was a resignation. In fact, he was left with no other option. This was a harsh blow for a man who had given so much of his life to public service. At age 57, with no desire for retirement and a considerable concern for his finances, Marshall wholeheartedly embraced the chairmanship of WANO. After a life in the public sector in the UK, Marshall embarked on a new career in private service covering the globe.[6]

****

When Marshall was unanimously elected Chairman of the WANO Governing Board in May 1989, it was evident that WANO had come a long way in a short time. In just 18 months, a group of senior utility executives had met in Paris, agreed on a need for a global organisation, and established WANO "to maximise the safety and reliability of the operation of nuclear power plants". Robert C Franklin, the head of Ontario Hydro and newly elected Chairman of the Board of WANO's Atlanta Centre, was awed by the accomplishment. "Anyone who has had any experience whatsoever in achieving international consensus will know just how difficult that has been and the accomplishments that these steps represent. These initiatives testify to the international spirit and cooperation that gave birth to WANO. Now we must make an honest effort to learn from the experience and expertise of others and to use that." Franklin warned that the new organisation was "only as strong as its weakest link. If one member country, one utility, should fall down on the job, then we will all fail." He challenged all WANO members to compromise, to be conciliatory, from time to time. Doing so would be "a sure sign that we are all contributing to and seeking excellence, the standard that we'll never compromise."[7]

Marshall's role in the formative days of WANO had, at times, been from a distance – most of his time was taken by his responsibilities at the CEGB. Although he had been supportive in assigning his close aide, Andrew Clarke, to the organisational planning staff, his personal involvement with the early activities of WANO was limited to encouraging the Planning Committee and keeping it to a schedule. Nevertheless, his individual success and public attention on larger stages, initially at the organisational meeting in Paris and followed by a greater triumph in Moscow, provided the enthusiasm and impetus for assuming the top WANO position. In addition, the opportunity to continue working in the nuclear safety field after the loss of his position at the CEGB convinced Marshall that WANO could best succeed with him at the helm. Subsequently, he threw himself into his job, visiting members throughout the world with Lady Ann during the early 1990s, and seizing on the idea of mutual self-help to improve the operations of nuclear power plants and restore the industry's damaged reputation.

As Chair of the WANO Governing Board, Marshall initially saw his role as the leader of a worldwide organisation whose choices would set the precedent for the level of collaboration between the four regional centres. Although his vision for the position and WANO would shift as circumstances changed, he remained optimistic and fully committed to improving the safety and operations of nuclear power plants and rescuing the industry's reputation from the damage inflicted by Chernobyl. At the Inaugural Meeting in Moscow in 1989, Marshall stated that he firmly believed in the importance of strengthening the cooperation and collaboration among nuclear operators across political and cultural barriers through four regional centres and a small coordinating centre "because we are anxious to avoid all bureaucracy. The whole idea is that we should learn from one another so that we emulate the best practices which are available throughout the world." Throughout Marshall's tenure, a high level of collective responsibility as preached by Marshall and WANO President Lee would be crucial in determining WANO's success or failure.[8]

After Moscow, Marshall's chairmanship began much as the creators of WANO had envisioned: a part-time Chairman who travelled to the regional centres as a booster for WANO programmes, to rally members to participate in operator-to-operator exchange visits, extol the value of exchanging operating experience through the use of Nuclear Network®, and encourage the identification and use of best practices. Marshall spent most of his time, which he estimated at 10% on behalf of WANO, on the road. But his role as Chairman changed radically in the late fall of 1989 as a result of the privatisation of the British utility industry. In November of that year, the Thatcher government surprised many utility executives – and especially Marshall – when it broke up CEGB, but did not privatise the country's nuclear plants. Marshall opposed the decision to exempt nuclear power, a decision made more onerous when a close associate and friend, John Collier, was named to head the new utility, Nuclear Electric Plc. The new situation, which left him in professional limbo, prompted Marshall's departure from the CEGB in December.[9]

****

The change in Marshall's employment status was just one of the transitions occurring at the time that affected WANO members. The arrangement put additional financial pressures on the regional centres, but especially Moscow Centre, which was beginning to experience a cash flow problem due to political and economic instability in the region. Not long after the founding of WANO, a series of protests in the summer of 1989 shook the foundations of the post-World War II political structure of the Communist states in Central and Eastern Europe and, eventually, the Soviet Union. In Poland, the Solidarity Movement under Lech Walesa ousted the Soviet-backed government. By the fall of 1989, Germans pulled down the Berlin Wall and began the process of reunification. Hungary, Czechoslovakia, Bulgaria and Romania all established non-Communist governments and, by 1991, the Soviet Union had dissolved, replaced by the Russian Federation. "The political changes in Eastern Europe created a new problem and a new opportunity for WANO which we did not anticipate when the organisation was set up," Executive Director Thomas Eckered recalled later. "The old reactors were all built by the Soviet Union to the standards of those early days. They are robust and easy to operate, but they are not as sophisticated as modern-day safety analysis would demand." How to move forward from Chernobyl became WANO's first major safety challenge.[10]

The sudden political changes in former Soviet Bloc nations between 1989 and 1991, particularly the reunification of Germany, opened the way for closer inspection of Soviet-designed and -built nuclear power plants by the West and a reassessment of their safety risks.[11]

This led to an appraisal of the safety of Soviet-built reactors by the IAEA and the countries of Western Europe, which shared a collaborative framework and common values regarding the safe operation of nuclear power. The West sought to impose safety upgrades on Chernobyl-type RBMKs – they lacked such safety features as containment buildings – and VVER-440/230s as they showed some safety deficiencies in their emergency core-cooling systems. The East German plant at Greifswald was particularly worrisome, as it had "deficiencies in safety technology…in nearly all areas investigated," according to a West German report. A nuclear expert from the Paris-based Organisation for Economic Cooperation and Development observed that the Greifswald units "are very far off our own regulations and requirements. Not marginally off, but incredibly far off." A West German news magazine claimed Greifswald was an "atomic bomb that could blow up any second".[12]

Problems elsewhere indicated that Greifswald was no exception. With the events of Chernobyl and the collapse of the Soviet Union, the West began to question the continued operation of the string of nuclear power plants along the spine of central Europe. The Iron Curtain described by British Prime Minister Winston Churchill in his famous 1946 speech at Westminster College in Missouri had been lifted to expose nuclear facilities using outdated technologies and lacking Western safety standards, in a line extending from Ignalina in the Baltic to Kozloduy on the Danube. Moreover, initial studies of the plants indicated that there was no standard solution for all plants.[13]

The Kozloduy plant in Bulgaria, about 200km north of Sofia on the Romanian border, was a particularly complex and thorny problem. It consisted of four first-generation VVER-440/230s and two newer VVER-1000 units supplying 40% of the nation's electricity, largely operated by a Russian staff. As economic chaos gripped Bulgaria, unpaid Russian workers left the site, followed by local operators who reportedly could make more money driving cabs in Sofia. An IAEA review team "found appalling housekeeping, significant fire hazards, ignorant and powerless inspectors and poorly trained operators." The sorry situation at Kozloduy was "due to a lack of safety awareness…and a bias toward production" that discouraged consideration of operational safety. "It would be imprudent," an IAEA nuclear safety expert concluded, "to continue to operate the plant until imperative repairs were made." The damning IAEA report seemed to confirm Western perceptions that the former Communist nations had defective reactors, unfit operators, deficient regulations and misguided management priorities. In the months after the collapse of the Soviet Union, a consensus hardened in the West, including among WANO members, that the Eastern European nations "could not be trusted with reactors of dubious safety".[14]

In addition to the VVER pressurised water reactors, the Chernobyl-type RBMK reactors – water-cooled, graphite-moderated units – also drew Western attention, though to a lesser degree than the VVERs because most were located farther east in the Soviet Union. Derived from the first Soviet power reactor built at Obninsk in 1954, the RBMK reactors were designed to be quickly built and easily maintained. Again, production of electricity, not operational safety, was the primary goal. A new generation of these plants had been installed in the 1970s and early 1980s at Leningrad, Kursk, Chernobyl, Smolensk and Ignalina in Lithuania. Located some distance from Western Europe, the RBMKs had operated satisfactorily and did not cause the same level of concern as the VVERs. But the fallout from Chernobyl intensified Western interest in the lack of containment and other safety features at RBMK units, as well as the mode of operations. Subsequently, the West urged design modifications in addition to the closure of the units at Chernobyl and Ignalina.[15]

While the Europeans planned a response to retrofit or close former Soviet Bloc VVER reactors, the US Department of Energy (DOE), long worried about the operation of Soviet civilian reactors, began an initiative in early 1990 to improve nuclear safety in the Soviet Union at the RBMK sites, dubbed *bezopasnost*. "We want to help the

Soviets develop an INPO-like approach toward improving plant safety because we believe they recognise the need to develop a culture of safe practices" similar to those developed by the Institute of Nuclear Power Operations (INPO) after Three Mile Island, a DOE spokesman explained. DOE contracted with INPO to run a two-phase project: first, to involve senior Soviet nuclear operation policymakers, then later to include plant managers, operations superintendents and training supervisors in improving reactor safety. INPO assistance included substituting symptom-oriented emergency procedures for event-oriented procedures; analysis of tools, skills and knowledge needed by operations and maintenance personnel; and development of performance indicators. Although many of these efforts were also part of WANO's programmes, the US government had the funding and the institution, INPO, to carry out the programme without WANO involvement.[16]

In any case, WANO's Charter prohibited such handling of government funds. "WANO will have nothing to do with governments, with regulators, with research or with commercial matters," Marshall had reminded delegates in Moscow. "Our sole concern is the safe management and operation of nuclear power plants throughout the world." Nevertheless, as Western Europe became increasing troubled with the operational safety of Soviet-designed reactors, pressure built on WANO to get into the game. In a speech to the Paris Centre Governing Board in Toledo, Spain, in the spring of 1990, Governor Werner Hlubek warned of the dangers of the Eastern European VVER reactors, particularly in their "deficiencies in design [and] in their management." He argued that the actions of the IAEA were not sufficient and called on WANO "to deliberate on whether and possibly how operators outside Eastern Europe could participate financially" in the retrofitting of the Soviet-built plants. "We should try to exert influence on the politicians," he maintained, convinced that WANO should "take the lead and should not leave action to others." Marshall, who attended the meeting, was sufficiently convinced to bring it before the WANO Governing Board a week later. The board agreed that WANO must address the problem.[17]

At the WANO Governing Board meeting in Moscow in July, Vitalii Konovalov, Minister for Nuclear Power and Industry, urged the Board to determine what help WANO could provide. The board decided to create a special organisation for this purpose. "We did not think this was a matter which should concern the entire WANO organisation and we therefore set up a special project which concerns only the Paris and Moscow Centres and which has extremely limited objectives," Eckered later wrote in a background report to the Governing Board on the history of WANO's involvement. Drawing on an IAEA technical analysis of the reactors, WANO hoped to define a solution "which obeys the ALARA principle; that is, a risk which is 'As Low As Reasonably Achievable'." WANO hoped that the Commission of European Communities (CEC), later known as the European Commission, or EC, would fund the work, but WANO was "not seeking [a] contract for this work and neither do we seek to make commercial decisions between contractors. We are not trying to replace the [contractors]; we are trying to help them reach a consensus judgement on what it required."[18]

WANO also wanted to make certain that there would be no opposition to its proposed VVER Special Project from the IAEA. Marshall wrote to Hans Blix, head of the IAEA, seeking a cooperative arrangement soon after the Moscow meeting. WANO officials met with Blix and his senior staff to further explain WANO's Special Project. The meeting, Eckered reported, "was very constructive and friendly and mutually supportive. The IAEA senior staff had already analysed that WANO and the IAEA were doing different jobs [and] that neither could attempt the work of the other."[19]

As a result of the discussions at the Moscow meeting, the WANO Governing Board created the VVER Special Project. It consisted of two expert groups, one from Moscow Centre and one from Paris Centre, headed by a Steering Committee chaired by Marshall. The Moscow group included Armen Abagyan, Alexandr Lapshin, the Deputy Minister for Atomic Power, and the Director General of the troubled Greifswald nuclear power plant. WANO would provide short-term "practical assistance to utilities in the USSR and Eastern Europe to improve the shortcomings of VVER plants". Jean-Pierre Baret, the technical resources director at EDF International, headed the Paris Centre group. To comply with WANO regulations, the new activity was to be carried out by a joint Moscow / Paris Centre subsidiary that would be part of WANO, "but operate outside the main mission of WANO." The subsidiary, not WANO, would channel funds from the CEC for the backfitting programme. The WANO Governing Board and Marshall agreed that the subsidiary "would be a one-off activity and not interfere with the prime task of WANO," though the expectation was that the project would take up to five years to complete.[20]

Not all WANO members were enthusiastic about this new WANO initiative. Zack Pate was uneasy about the new direction WANO was taking with the VVERs. While he was in favour of the concept of retrofitting the plants, he stated that WANO's involvement in a consultative capacity or as an agent went further than its mission, and effectively did "stretch the Charter". He explained that neither the Atlanta nor the Tokyo Centres should be involved with European Communities funds, though he would accede to the Paris and Moscow Centres forming "a joint subsidiary to specifically handle this initiative." Lee agreed, hoping the subsidiary "would help to avoid any diversion from the established WANO mission". With those strictures in mind, the Governing Board approved the initiative to "provide practical assistance to utilities in the USSR and Eastern Europe to improve the shortcomings on VVER plants."[21]

Throughout the fall and winter of 1990–1991, the two expert groups met to decide on a course of action related to the design and operation of the VVER reactors. Working from a paper drafted by the Moscow Centre group, the team sought to create a list of actions that would make the safety of the VVER-440/230 reactors acceptable and permit their operation for another three to five years. The Moscow/Paris team concluded that backfitting might aid in correcting some design flaws, but an equally serious issue – fixing the slack safety culture of plant operators and managers – demanded considerable WANO assistance. The broad agreement reached between the Moscow and Paris Centres was an important precedent for WANO. "We can certainly expect that this way of working together will continue in the future and achieve a full understanding of two different cultures," the Paris Centre director reported optimistically.[22]

As Chair of the Steering Committee, Marshall became increasingly involved in the special backfitting programme, which some governors thought came at the expense of WANO core programmes. In addition, he also became heavily invested in another major VVER project – an IAEA effort to funnel assistance to Bulgaria's Kozloduy power station. Kozloduy became the West's poster child for reactor assistance when, in the fall of 1990, the European Economic Community designated significant funding to upgrade the plant. Pate and the Atlanta Centre staff worried that Marshall's personal interests were drawing WANO "into extensive involvements" with governments and commercial matters, such as the "distribution or administration of government funding of nuclear plant improvements," to the detriment of the organisation.[23]

Nevertheless, by the summer of 1991, Marshall focused on Kozloduy. He also visited a number of the VVER-440/230 reactor plants, all of which had "generic issues that were well defined". Kozloduy, however, was a "special case," he cautioned the VVER Special Project Steering Committee. "In addition to the generic issues, they have unique, specific needs that have to be addressed in an urgent manner," he told WANO's VVER Advisory Group. The needs, which pertained to twinning and management, or "housekeeping" problems at the plant, were so great that "they could not be adequately addressed by the Special Project in its present form," according to Marshall. The problems at Kozloduy were only partly technical. "There are considerable shortcomings both in the morale of the staff and the organisation and management of the plant," a WANO report noted. The Advisory Board agreed and ended its support for backfitting in favour of Marshall's new emphasis on sponsorship of the Kozloduy Twinning Programme and management reforms, both of which would be underwritten by the CEC. Over the next few years, the Bulgarians operated two units at Kozloduy while upgrading the safety features of the remaining two. The Bulgarian government raised salaries for the plant workers, and many of the experienced operators returned. The plant continued to produce electricity, and WANO pushed both Bulgarian regulators and Kozloduy management toward a more safety-oriented culture.[24]

The tensions between the Americans' focus on WANO's core safety programmes and the Western European emphasis on upgrading Soviet-built reactors, which Marshall favoured, appeared more serious to outsiders than they actually were within WANO. The Americans based their position on the successes of INPO. Marshall had a broader and more complex constituency to consider. He was sorting out priorities for what a new international organisation could do to establish a viable role among the majority of its members. While the Americans differed with Marshall and the Europeans on how WANO should proceed, both sides, importantly, agreed that the old nuclear order in Eastern Europe had to be supplanted and that WANO was the organisation to accomplish the job.

There were no differences regarding the importance of WANO's programme of exchange visits between Moscow Centre plants and utilities from the other three centres. By December 1990, 45 of 50 exchange visits had been completed in spite of visa issues or postponements. For an association just beginning to co-operate internationally after years of isolation and mistrust, the organisation of the technical and social aspects of the visits and the willingness for detailed discussions of technical matters were "without exception, very good and appreciated", Marshall reported. Moreover, a "number of operators expressed their interest in more intensive exchanges of operating practices, personnel, technical documents and other means of making their operational experience valuable to their Technical Exchange Visit partner and to other operators." Some plants, such as Catawba in the US and Zaporozhye in Ukraine, entered into long-term exchange agreements. One WANO concern with the exchange visits was the reporting of the visiting teams. The reports lacked uniformity and "could be more technically detailed in some cases," the head of the programme explained. Good reporting of the visits could "help stimulate other operators in preparing visits in the future." At WANO's insistence, the visits continued.[25]

Gathering consistent reports proved to be a nagging WANO problem elsewhere as well. The Operating Experience Information Exchange programme, established to report significant events, remained unsatisfactory. Language differences were the major, and toughest, problem to overcome. Another issue was that a number of members had yet to be connected through Nuclear Network®, and technical shortcomings remained in other areas. After analysing the programme, Eckered reported that the number of events was "too low" and only a small percentage of those were reported to the regulatory authorities. Experience at INPO had demonstrated that even events that seemed minor at the time, when put together, were important in indicating generic weaknesses or common malfunction trends. The WANO Governing Board agreed to sponsor a workshop with the goal of better explaining the necessity and use of event reporting, though Rémy Carle questioned the programme's value if operators did not make use of it. Carle's query went to the heart of the matter. The fear of repercussions from reporting a minor accident, the different cultural responses to operational errors or "failure," the linguistic tangles of explaining technical events, and connecting to Nuclear Network® all worked against the success of the information exchange programme and would continue to frustrate WANO's goals for the programme.[26]

****

Amid the geopolitical and economic changes in the Soviet Union and Eastern Europe, WANO held its first Biennial General Meeting (BGM) in Atlanta, Georgia, in April 1991 at the Waverly Hotel, not far from WANO's offices. More than 240 delegates representing 27 nations attended. Marshall reported on the progress of the association over the previous two years, and members had an opportunity to express their views and concerns on how WANO had functioned, how it might improve and what direction it might go in the future. "There was a clear consensus that WANO was on course to achieve its goals," Marshall reported, but reaching those objectives required more member commitment and participation in the programmes. One goal, however, was attained soon after the meeting: the Romanian Power Authority (RENEL) became a member of WANO. Thus, by the fall of 1991, every nuclear utility in the world had joined WANO and was involved with its work.[27]

The Atlanta meeting was an opportunity for WANO's Atlanta Centre to showcase North America's strong embrace of nuclear safety through INPO, in whose offices the WANO centre was housed. Lee, who was stepping down as WANO's first President, was honoured not only for his commitment to WANO and "his inspiration and guidance during the formative years of WANO" but also for his leadership in creating INPO and a vigorous safety culture in the US. A number of the American hosts firmly believed that the INPO example held great value for WANO and hoped that utility executives and workers in other countries would learn from and adopt INPO's example. To replace Lee, the delegates unanimously elected Shoh Nasu, a former president of the Tokyo Electric Power Company (TEPCO), as the next President of WANO and host of the second BGM to be held in Tokyo in 1993.[28]

The Atlanta BGM was the first time many officials from the Soviet Union and Eastern Bloc nations had been allowed to travel freely and visit the US. For many, Atlanta meant an outstanding opportunity to shop, and the meeting's organisers planned a schedule for attendees' spouses with that in mind. Buses fanned out from the Waverly to Atlanta's popular shopping malls. For storekeepers and shoppers alike, it was Christmas in April. Nevertheless, planners did not ignore Atlanta's cultural attractions. Pate's wife, Bettye, made certain that spouses could visit the Atlanta Historical Society and the Swan House, as well as take a private tour of the "new Southern estate" of Mrs Deen Day Smith, which was filled with antiques from around the world. The social programme also included tours of Atlanta, the Carter Presidential Center, the CNN Center, the Atlanta Botanical Garden and the High Museum of Art. On the final evening, delegates and guests banqueted at a noted downtown restaurant, tapped their feet to the Peachtree Strutters jazz group and listened to the Atlanta Pops Orchestra. To end the evening, they toured Underground Atlanta and wandered through the Coca-Cola Pavilion, home of Atlanta's best-known export.[29]

By the summer of 1991, Marshall was spending more than 90% of his time working on Kozloduy and the Special Project regarding other VVER reactors in the East. Marshall's level of participation continued to cause unease in Atlanta and London. "I am distressed that you are having a series of discussions with US government officials regarding WANO," Pate wrote to Marshall in May 1991. Pate had learned that

Marshall was lobbying a senator to include WANO in a proposal that would have the US government match European funding to improve Soviet-designed reactors and sought to nip the initiative in the bud. On behalf of the Atlanta Centre's Governing Board, Pate warned Marshall not to engage in discussions that would "result in any WANO involvement, direct or indirect, with US government financial aid intended for nuclear improvements".[30]

Pate worried that Marshall's intensive commitment in the VVER retrofitting programme and Kozloduy, both of which benefitted about 10 nuclear units, was coming at the expense of the broader WANO core programmes, which were aimed at assisting more than 400 units: "The signal that this is clearly sending is that those special projects are the primary business of WANO, rather than the mainstream programmes that are spelled out in our Charter and our long-term goals." Pate complained that Marshall's focus on other projects was not leaving sufficient time at Governing Board meetings to adequately discuss WANO's budget and core planning goals and objectives. The regional directors, Pate argued, "had spent a lot of time preparing them and we owe them more of our time and attention". Due to Marshall's activity, Pate said, WANO must be careful to carry out the mandate of its membership, especially during the association's formative years "when the credibility we build will be our most important asset in the future". As it was, WANO was "losing credibility with many Atlanta Centre members and with key elements of the US government. That trend," he warned Marshall, "must change."[31]

Unspoken in Pate's opposition to Marshall's initiatives was the fact that Marshall had trod on INPO's turf. During a trip to the US that spring, Marshall had spoken to Admiral James Watkins, Secretary of the Department of Energy, asking if Bulgaria and Czechoslovakia could participate in a DOE-sponsored project providing training and operating procedures to the Soviets. It did not take long for word of the meeting to get to Pate, to whom Marshall had said nothing of the visit. INPO had a contract with the US government to provide assistance with the reactors in central Europe. Thus, Marshall's conversation with American officials such as Watkins, who believed that the safety issues of Eastern European reactors were damaging the American nuclear power industry, was moving in on INPO's turf in addition to being outside the boundaries of the WANO Charter.

The Americans' concern that the Special Projects had occupied much of Marshall's time was somewhat accurate. Marshall had indeed expanded his role beyond that of the WANO Charter at a time when worries about precedents loomed large in the minds of many of WANO's incorporators. As Marshall and WANO withdrew from an active role in retrofitting VVERs in the former Soviet Union and provided assistance to Kozloduy, the fledgling organisation, in retrospect, may have missed an opportunity to push incentives for operational safety to the European Union or, at the least, have significant influence in international safety discussions. As an international collaborative framework evolved from the Western nations' efforts with reactors in Eastern and central Europe, the result was, to many, disappointing. The Convention on Nuclear Safety, adopted in 1994 and based largely on IAEA standards, lacked incentives, and there were no verification or enforcement mechanisms. It was not until the European Commission issued *Agenda 2000*, a roadmap for EU enlargement, that the West had sufficient leverage to demand that former Soviet Bloc nations meet Western nuclear safety standards as a condition for admission. As WANO's Charter limited direct involvement with governments, it could only play a lesser direct role than Marshall would have liked in this aspect of improving nuclear safety.[32]

To be fair to Marshall, he was extremely sympathetic and sensitive to the needs of the central European nuclear plants. He was also responding to a pointed concern among many WANO members to allow an expansion of the WANO Charter to permit the chairman such direct involvement, or so Marshall interpreted that sentiment. Such a revision of the Charter might well have positioned the organisation to respond more flexibly in the future with more direction from the chair. But direct involvement with governments, with one notable exception, would remain outside the Charter.

What WANO could accomplish within its Charter and the framework established by the IAEA was a backfitting programme of hardware and software advances which, when implemented, would "lead to substantial improvements in the safety of VVER-440/230 reactors" as well as technical aid. But the Governing Board halted Marshall's retrofitting initiative and further cautioned him not to intrude on the IAEA, to which the CEC had given the leadership role in dealing with the Bulgarians and Kozloduy. While WANO's role was not all he had hoped for, Marshall made Kozloduy its highest priority. WANO held three contracts to upgrade the operational safety at Kozloduy that comprised a twinning arrangement with Bugey, a Paris Centre plant, to exchange information and technical staff, an outage assistance team of engineers to restore the operations at the plant and assistance to backfit the hardware projects. But the Governing Board, led by Carle, insisted that backfitting be handled by others and that even the "urgent management issues" of Kozloduy "be clearly defined in scope and nature" before giving its approval. By 1993, WANO's role was "entirely to advise and assist the Bulgarian operator to implement safety improvement programmes".[33]

If frustrated in not playing a larger role in the retrofitting of the reactors in the former Soviet Union and its satellite countries in Central and Eastern Europe, Marshall eventually found WANO's voluntary pilot Peer Review programme to be an exciting alternative. WANO's first President, Bill Lee, had pushed hard for peer reviews, arguing that they had been enormously successful for INPO. Bill Cavanaugh and Zack Pate, Atlanta Centre's representatives to the WANO Governing Board, also favoured an international Peer Review programme. Marshall, however, had taken a European view of peer reviews, otherwise termed "technical reviews" or "audits", and did not push for them, favouring instead the twinning arrangements between plants such as that between Kozloduy and EDF's Bugey station. But in the spring of 1991, while visiting the US, Marshall saw for the first time how an INPO peer review was done.[34]

At INPO's initiative, Wesley von Schack, President of Duquesne Power, invited Marshall to participate in the peer review at the Beaver Valley nuclear power plant near Shippingport, Pennsylvania, also the site of the country's first commercial nuclear plant, which had been decommissioned in 1982. While accommodations at the La Quinta Hotel near the Pittsburgh airport were hardly the calibre to which Marshall and Lady Ann were accustomed, von Schack recalled how engaged Marshall became during the two-week long on-site review. He worked with the evaluation team, ate dinner with them and spent evenings around the bar at the hotel learning how the peer review process unfolded. Von Schack was impressed by Marshall's "roll-up-the-sleeves" approach to peer reviews to the point of placing a fold up bed at the plant on which Marshall could take naps during the visit. Von Schack felt some guilt over putting up an English lord in a chain motel, so he hosted them for a weekend in the Mellon family cottage at the exclusive Rolling Rock Club east of Pittsburgh. It was an expensive week, von Schack later recalled, but the visit accomplished what Pate had intended. Thereafter, Marshall became a convert to peer reviews. His backing of peer reviews earned him high marks from Atlanta Centre and eased the tension that had developed over his involvement with Kozloduy.[35]

Even so, the Governing Board's support of peer reviews remained mixed – there would be no rush to judgement. Although the concept was backed by Paris Centre, Carle believed that peer reviews should be done on a voluntary basis and that each country that carried out such evaluations should report on its experiences with regard to the usefulness of the reviews and how they were conducted. Adolfo de

Ubieta from Spain believed that the reviews could "be a powerful tool to close the operating experience feedback loop" but indicated that Spanish utilities were not "institutionally prepared" to conduct them. The director of Moscow Centre thought peer reviews were a good idea but declared that it was "premature in the context of Eastern European plants" and that it would "take some time to build up the necessary confidence to achieve the openness required". Members from Tokyo Centre worried that evaluations might be harmful if the results were not properly restricted. India conducted what it termed "introspective" reviews. The Japanese allowed that an international approach would be valuable "provided it was attempted slowly and on a friendly basis. The cultural questions," Ryo Ikegame, a member of the Tokyo Centre Governing Board cautioned, "should not be underestimated", meaning that peer reviews in Asia would not conform to Western standards. According to Tokyo Centre, one size for a peer review did not fit all plants; peer reviews would need to be tailored to each situation.[36]

At the Atlanta BGM, Lee, Pate and Marshall vigorously pushed to make peer reviews "part of the forward-looking programme for WANO". The Governing Board remained cautious "in order to account for cultural differences and national practices". The Governors agreed to initiate a pilot voluntary peer review programme that, though led by Atlanta Centre, would lack any fixed implementation procedure. The programme consisted of a series of pilot peer reviews to be held at a number of "volunteer plants" before the next BGM in Tokyo in 1993. The pilot Peer Review programme was to test the procedures and assess the benefits of peer reviews, which would investigate the organisation, operations and practices of a nuclear power plant and note its strengths as well as areas in which improvement could be made. At the end of the review, the team would provide recommendations for improvement. The distribution of the final report was not resolved, though the intent was to keep the reports confidential within policy guidelines to be established by the WANO Governing Board. Peer reviews would serve two functions. The immediate objective was to improve safety and reliability at the plant. In addition, the programme also sought to train team members from the various regional centres in conducting future reviews, as there was no tradition of peer review among WANO members outside North America and, to a lesser degree, France. The Paks plant in Hungary offered to be the first to host a peer review team consisting of experts from all the regions.[37]

Although WANO's Governing Board approached the pilot programme gingerly, the Americans, convinced of the effectiveness of peer reviews, believed it would become a major WANO programme by the end of 1993. In a July 1991 memo entitled *Desired Outcome, Pilot Peer Reviews*, Walter J Coakley, the Acting Director of the Coordinating Centre, wrote to Stan Anderson at Atlanta Centre that "we expect that by 1993 all regional centres will commence a peer review programme". The final programme "will be most effective if it resembles to the extent feasible the methods we [at INPO] have developed". Yet "any attempt to impose our methods on the Peer Review programme directly would be met with strong opposition. Our goal then should be to adapt our methods to the situation posed by the pilot Peer Review programme, package them in special documents, and use them to conduct productive visits." If the Peer Review programme were to go forward as a separate programme in 1993, Coakley argued that "there must be strong endorsements" from the management of the utilities receiving the site visits. To achieve that, he advised that the host plants see "real value as well as perceived benefits from the reports. Acceptance will be greater if the teams can find some valuable strength that can be published during 1992 to WANO membership." In short, Coakley thought WANO should adopt a peer review programme that emulated INPO's evaluation process "to a significant extent but is flexible enough to satisfy the autonomous views of the regional centres." The programme was given an extra boost when

Coakley replaced Eckered as the Acting Director of the London Coordinating Centre at the end of 1991.[38]

The Paks plant in central Hungary consisted of four second-generation VVER reactors, 440/213s, that had been designed to include the containment buildings and emergency core cooling and auxiliary feed water systems lacking in the older 230 models. The units were relatively new, having come on-line during the mid-1980s. When WANO informed the IAEA staff of the pilot Peer Review programme in September 1991, the IAEA was "unhappy" claiming the reviews would only duplicate the IAEA's OSART (Operational Safety Review Team) programme. "We tried to convince them that this was not true," Eckered reported of the encounter, "but only our promise to meet with them directly after the first peer review [and] inform them about our experiences could lighten the atmosphere somewhat." By November the situation with the IAEA was defused, the IAEA concluding that both approaches were "valuable and complementary". Plans went ahead for the first pilot review at Paks, which occurred over two weeks in February 1992. By the end of 1993, seven more pilot peer reviews took place: at Diablo Canyon in the US; Bruce nuclear generating station in Canada; Koeberg nuclear power station in South Africa; Chinsan nuclear power plant on Taiwan; Angra 1 nuclear plant in Brazil; Balakovo nuclear power plant in Russia; and Tomari nuclear power station in Japan.[39] Pate hand-picked the team leader for all of the pilots except Tomari, and he personally attended the last few days and the exit meeting for Paks, Balakovo and Koeberg.

While the Special Projects and peer reviews occupied much of Marshall's time, WANO established precedents for operating its principal programmes that were fundamental to the association's mission, such as the exchange of operating experience through the Nuclear Network® computerised messaging system, operator-to-operator exchange programme, plant performance indicator programme and good practices programme, as well as a series of workshops and seminars designed to increase the sharing of operating experience between plant personnel.[40]

Based on the INPO experience, there was much hope that the information exchange of operating experience would become a valuable weapon in WANO's safety improvement arsenal. Staff in London carefully crafted specific criteria for nuclear plants to use in reporting events, so that the report content would be analysed and distributed to operators around the world. The event reports, WANO officials hoped, would provide "the opportunity for each plant to examine its operation and, where applicable, take steps to preclude a similar problem from occurring". But even with upgrades to Nuclear Network® to simplify reporting and improvements in the quality of event report content, most plants outside the US were reluctant to report operating deficiencies; consequently, the number of event reports did not meet the "lessons learned" expectations of the WANO Coordinating Centre and Atlanta Centre. The Governing Board, led by Carle and Hlubek, believed that valuable opportunities to share experiences were being lost. The programme did, however, accumulate an extensive data bank of operating experience, even if its utilisation was disappointing. Nevertheless, by the end of 1993, the Operating Experience programme remained very much a work in progress.[41]

Far more successful was the programme of exchange visits between plant operators and the seminars and workshops that gave nuclear personnel an opportunity to meet, discuss and learn through direct dialogue, particularly among the operators in the West and those from the former Soviet Union and Eastern Europe. Vladimir Fuks, a governor from Moscow Centre who ran a nuclear power plant in Ukraine, told the Governing Board in 1992 that "the real value of WANO was in assisting operators to improve the quality of plant operation and performance, against unhelpful and critical advice of other [unnamed] organisations and institutions." Marshall admitted, however, that "there is still much to do if the best professional operating standards are to be achieved across the world."[42]

The Good Practices programme had languished on several fronts. Many of the WANO Governors believed the problem lay in the fact that the practices were not fully identified and communicated to the membership. The lack of acceptance in individual plants stalled the adoption of good practices. The Governors proposed that introducing the topic into workshops and seminars would encourage individual plants to adopt them. Marshall suggested singling out good practices at top operating plants, rewarding excellence as INPO did, but the Governing Board blocked such a move, citing the "national sensitivities involved". The Board also labelled a report on good practices from Tokyo Centre as "complex" and needing "further consideration". Hung up on identification and application issues, as well as a lack of widespread member interest, the Good Practices programme remained in limbo.[43]

Although events forced Marshall's initial vision for WANO to vary from the original script, with the Governing Board's approval he adroitly switched course to take advantage of opportunities – such as the VVER Special Projects programme and peer reviews – as they arose. In doing so, he involved the Paris and Moscow Centres into the core of WANO activities, expanded the Charter accordingly and gave WANO a higher profile internationally than it might have otherwise experienced in concentrating solely on its core programmes. Of course, Marshall did not hold together this new multinational organisation by himself. The experience of INPO, the extraordinary work of the Coordinating Centre staff, the contributions of EDF and the crucial participation of the Soviet Union all ensured the success of an international association committed to nuclear safety. Nevertheless, Marshall was the glue that held together the confederation of regional centres under the WANO banner in its formative years, kept critical utility leaders involved, established its worldwide credibility and provided a foundation on which the association could prosper. If Marshall had once scoffed at WANO's potential, he was largely responsible for shaping a flourishing international organisation to promote and enhance nuclear safety.

****

Marshall did not look forward to stepping down from his WANO chair. There was some discussion that he could be elected President of WANO, but that idea received little support beyond Marshall's friends in Japan. Rather, Pate worked out an alternative plan, one that recognised Marshall's considerable contributions to WANO while making use of the great reservoir of goodwill and the respect with which he was held by WANO's members. The position, as Pate envisioned it, recognised Marshall's "skill and the great value of [his] visits to countries operating or wishing to operate nuclear power plants". The Governing Board wanted Marshall to continue these visits in a new role – as "Special Ambassador for WANO". Importantly, the position came with a "Special Assistance Fund" of £50,000, controlled by the London Coordinating Centre, to allow Marshall to "undertake specific assignments on behalf of WANO". But the Special Ambassador was put on a short leash. With Marshall's tendency to ignore budgets, regional chairmen who would foot the bill wanted to ensure that "costs are carefully evaluated and monitored. And expense reports meet audit standards."[44]

While Pate was creating the position of WANO Special Ambassador for Marshall, the Chairman's Russian friends were drawing up a plan that would also put his services to use. Relations between Russia and Ukraine had been strained since the collapse of the Soviet Union. These political divisions had by and large ended effective cooperation between the utilities in the two countries. However, the newly appointed Russian Minister of Atomic Energy (Minatom), Viktor Mikhailov, and President Leonid Kravchuk of Ukraine both saw the need for assistance at their troubled nuclear power plants and agreed in principle to a "Users' Group" headed by Marshall. They believed that a non-governmental organisation led by someone of international stature might be able to circumvent the political differences "preventing cooperation within the states of the former Soviet Union" and coordinate the needs of the various utilities in Russia and Eastern Europe. Abagyan, Eric Pozdyshev, who was the head of Concern Rosenergoatom – which governed Russian nuclear plant operations – Hlubek and Carle backed the idea for the prospective Special Ambassador to bring the utilities together in the face of governmental estrangement. When the Governing Board began discussing the devil's details rather than the concept itself, Fuks, who knew firsthand the problems of running a nuclear power plant in Ukraine, stopped them. He reminded the members that the situation was critical and "only the influence of WANO stood any chance of bridging the political divisions." That was sufficient for the Governing Board to support the plan for Marshall to assist Moscow Centre utilities "in working together to achieve common objectives in the field of operational safety".[45]

In his four years as Chairman, Marshall had accomplished much. The fledgling organisation that Marshall had been certain would not survive had thrived under his leadership. WANO firmly established itself as an effective international nuclear safety leader. Although Marshall claimed, with some exaggeration, that the four programmes that formed the core of WANO's activities – the exchange of information on significant events, exchange visits between plants, the identification and dissemination of good practices and measuring plant performance indicators – were "well established", they at least had begun to function. The exchange visits were the most noteworthy and successful, clearly demonstrating the value of direct dialogue between plant operators. The Operating Experience programme had a patchy reporting record, and most of the data submitted came from American plants. Judgement was still out for the Performance Indicator programme, as WANO had only recently distributed a set of measurement criteria on "which operational excellence will ultimately depend". However, the Good Practices programme, assigned in 1989 to Tokyo Centre, had not worked as designed. An analysis of the Governing Board in the fall of 1992 determined that "good practices were still not fully being identified and disseminated to the membership." As a result, "too many opportunities were being lost."[46]

As he turned over the chairmanship to Rémy Carle in the spring of 1993, Marshall reflected on WANO's successful evolution during a time of momentous political and economic upheaval. At the top of his list was the support WANO provided to the utilities of Eastern Europe that were affected by the breakup of the Soviet Union and the political upheavals that followed throughout the former Soviet Bloc nations, particularly Kozloduy. "These tasks are far from easy," he wrote. Technical complexities and the vagaries of governmental and international bodies added to the difficulties. In addition, "vested interests, lack of urgency and shortage of resources have all hampered the work, but progress is being made. I am proud to have been involved in this activity on behalf of WANO…and look forward to continuing this association in the future."[47]

Marshall also praised WANO for the enthusiasm the utilities demonstrated toward the pilot Peer Review programme. While the value of peer reviews had been proven for a decade in the US, Marshall was impressed with how the reviews had been embraced by WANO's members – far beyond the expectations of the organisation's founders. Led by Atlanta Centre, the translation of peer reviews "across cultures and national boundaries and the willingness of so many plants worldwide to undergo review," he commented with pride, "was in my opinion a most remarkable demonstration of the mutual confidence and respect that exists between our members."[48]

Marshall's feeling was fully reciprocated by the WANO Governing Board. He had made "a unique contribution to the formation and development of WANO and has served the Association with distinction as its first Chairman from 1989–1993. By his foresight and effective leadership during its formative years, WANO has now established a position of strength and influence to the benefit of its members in their pursuit of excellence in matters of operational safety." The Governing Board's resolution in thanking Marshall emphasised his "inspiration and guidance and untiring efforts", which had earned him "the admiration and respect of utilities worldwide". The Governors specifically honoured his dedicated service and work to develop "effective and worthwhile" exchanges between operators in the former Soviet Union and Eastern Europe, and those of the rest of the world. Of course, the resolution papered over the occasional tensions that Marshall's chairmanship had experienced, but there was no question about the great respect and affection that WANO members had for him. WANO's first Chairman, who had done so much to hold together all the WANO programmes and parts, would continue as Special Ambassador. He would be stepping down but, thankfully in the eyes of most, not away.[49]

Chapter Three

PROTECTING **THE CORE**

Although not widely recognised at the time, the election of a new Chairman, Rémy Carle, at the Tokyo Biennial General Meeting in April 1993 allowed WANO to refocus on the core programmes upon which it had been founded, plus the Peer Review programme. As long as Lord Walter Marshall was both the WANO Chairman and the head of the VVER Steering Committee and Users' Group, the Governing Board found itself stretching WANO's financial resources and splitting the chairman's time between two tasks – one toward its core programmes and the second toward its special role in assisting Eastern Europe. While many members considered Marshall critical to the success of WANO's Special Projects in Eastern Europe, several of the board members, led by those in Atlanta Centre, worried that the core programmes were being neglected in favour of the retrofitting and safety programmes directed at Soviet-designed power plants. Now, with Marshall stepping down from the chairmanship, Zack Pate recommended that he become a Special Ambassador to those efforts assisting with nuclear plants in the former Soviet Union. With that move, Carle could refocus on WANO's core programmes while Marshall served as WANO's Special Ambassador working with troubled Eastern and central European nuclear plants. As an initial part of his refocusing effort, Carle pushed for a review of WANO's first five years and an analysis of what the international voluntary organisation of utilities had achieved – or not achieved – since its founding.[1]

Carle's election, however, was not without controversy. Marshall's friends on the WANO Governing Board from Russia and Japan urged that the Articles of Association be changed so that Marshall could continue as chairman. Atlanta and Paris Centre governors, led by Pate, William Cavanaugh III and Carle, strongly disagreed, partly because they did not want to set a new precedent. In addition, while they believed Marshall had done an excellent job in launching WANO, they did not support his personal emphasis on the Soviet-designed reactors if it in any manner was at the expense of developing the organisation for the benefit of all members. In short, they did not believe WANO could move forward if Marshall continued at the helm. The Russians, led by Armen Abagyan and Eric Pozdyshev, argued that if Marshall did not continue, they might withdraw. A Tokyo Centre governor, Ryo Ikegame, sided with the Russians.[2]

The impasse was felt throughout the association. The chilly relations filtered down to the regional boards. On top of it all, Eckered had left as the director of the Coordinating Centre. The impasse was finally broken by Eckered's replacement, Andrew Clarke, who, with Pate's strong backing, brokered a deal that would retain Marshall as WANO's Special Ambassador assisting the Eastern and central European nuclear plants with funding from INPO and Électricité de France. The Russians and the Japanese agreed to the compromise. Marshall, who was initially cool to the idea, finally accepted the concept on the premise that WANO might be severely impaired if those programmes failed. The deal opened the way for Carle's unanimous election in the spring of 1993.[3]

WANO's second chairman had an impressive CV, though his career focused on the application of nuclear energy in the business side of the industry rather than the basic research that had been the hallmark of Marshall's career. Born in Paris in 1930, Carle received a classical education of humanities and Latin and Greek. At age 17 he switched to science and engineering, enrolling in mathematics and technological studies. In 1953 he graduated with an engineering degree from the École Polytechnique in Paris and received an advanced degree from the École Nationale Supérieure des Mines de Paris the following year. In spite of its name, the École Nationale Supérieure des Mines de Paris trained leaders for careers in heavy industries like steel or energy more than coal mines. Carle became interested in managerial and economic issues, especially the industrialisation of technology. He joined the Commissariat à l'Énergie Atomique (CEA) as an engineer in 1955. He was involved with nuclear research and development of advanced reactors, including the first industrial reactor in France at Marcoule, whose construction he coordinated. Extensive nuclear construction experience gained over the next few years provided him with a position that helped define his entire career: the building of the Phénix prototype reactor. Carle had proposed building the plant as a joint venture with a team consisting of CEA and Électricité de France (EDF). The team completed the plant within budget and on time. Carle became CEA's Director of Reactor Construction. He also served as Director of the Centre d'Economie Industrielle (CERNA), a research laboratory of industrial economics and finance, and founding President and Chairman of TECHNICATOME, the division of CEA that controlled the work on nuclear reactors used in military applications such as submarines and aircraft carriers.[4]

Because of his extensive contact with people at EDF through the Phénix project, in 1976 the company invited him to join the organisation. He became Director General of its Engineering and Construction Division in 1982, responsible for the company's Nuclear Power Construction programme. He was appointed Deputy Manager of EDF in 1987. Carle had been crucial to the success of the Paris meeting in 1987 and had continued his strong involvement with WANO as a member of the Paris Centre Governing Board since that time. According to some, Carle and Pierre Tanguy, the representative to INPO's International Participant Advisory Committee, were the driving force behind EDF's backing of WANO. A highly respected and admired utility executive, Carle became the public face of EDF after the Paris meeting. Pate, who applauded Carle's election, considered him a statesman, a trait that would be put to the test as he sought to revitalise WANO's membership and programmes.[5]

Carle was more than a nuclear engineer. Since childhood, Carle loved music, and he played the piano, harpsichord and organ. An accomplished pianist, he had a passion for classical music, especially the piano, cello and violin trios of Schubert, which he enjoyed playing with his son-in-law and a friend. He was also an author, writing *L'Électricité Nucléaire* with Michel Durr. Published in 1993, the book was translated into English as *Nuclear Power*. Tall, with thinning hair and an open face given to broad smiles, Carle had a reputation for dressing more like an engineer than a French businessman – neat but not natty. He was fluent in English and a passionate supporter of nuclear energy and WANO. Importantly, the industry viewed him as someone who could get things done. "He didn't accept no for an answer," recalled Andrew Clarke, who worked closely with him at the time, "and people found it quite difficult to say no to him because he had credibility." Moreover, he thoroughly understood the cultural and political challenges WANO had to overcome. He had befriended Zack Pate during their work on WANO's Budget Committee, and the relationship was strengthened by their common view on WANO's proper course – the improvement of its basic programmes. As WANO approached its fifth anniversary, Carle wanted the Governors and centre directors to analyse the effectiveness of WANO, review its structure and activities, and identify any shortcomings.[6]

As a member of the Paris Centre Governing Board, Carle had become frustrated with what he viewed as the ineffectiveness of the regional governors and the declining commitment among many WANO members. After four years, many of the original WANO leaders were leaving the regional boards and the Governing

Board. They were not being replaced in some regions by men of like stature. For example, when Werner Hlubek left the Governing Board, German involvement in WANO dropped significantly. As a result, Carle believed that the value of WANO to members in all centres with the exception of Atlanta was diminished. In addition, the centres lagged in producing events reports, which he believed were fundamental to WANO's long-range success. But the issue was extremely sensitive – partly because of language difficulties and a reluctance to admit to errors, partly because each region believed it could best implement WANO programmes as it saw best and partly because many members considered that their own safety policies were superior to others and saw limited value in WANO's programmes – so Carle and the Governing Board tiptoed around it, only suggesting that "all centres should give the matter serious consideration".[7]

As Chairman, Carle continually sought to get WANO and its members to work more effectively. To do this, he visited as many places as possible and spoke to planning staff and utility executives to get them to get on board with WANO's programmes. He toured members' plants throughout the world, including Angra in Brazil, Embalse in Argentina, Point Lepreau in Canada, Balakovo and Kursk in Russia, Rovno (Rivne) in Ukraine, Kakrapar in India and KANNUP in Pakistan, as well as the inauguration of the Daya Bay plant in China, which EDF and Carle helped build. Such personal visits were essential for the chairman to get the pulse of WANO's members and programmes. While he had been "well received" at the plants, not all drew on WANO's resources or participated in WANO's programmes, he reported to the Governing Board. Carle wanted to change this attitude. Moreover, the failure of India and Pakistan to sign the Non-Proliferation Treaty restricted outside assistance. Both Canada and Japan prevented their nationals from assisting the two nations. Carle confessed that WANO, too, was limited in what it could do. He believed that WANO could help "by keeping in close touch" with the two countries and alerting them "to opportunities to make progress without infringing political constraints." The tour allowed Carle to see both the strengths and weaknesses of WANO and what might be done.[8]

The new WANO President, Ian C McRae, who was CEO of the South African national electric utility Eskom, set out to reconnoiter the WANO centres and learn more about the working of the organisation. Because of their government's policy of apartheid, South Africans found their travel circumscribed by nations that found the policy unacceptable. McRae had trouble travelling to the US and Eskom's Koeberg plant had been blocked from joining Atlanta Centre because of it. A quiet opponent of his government's racial separation policy, McRae positioned Eskom to be a leader in meeting the social, economic, and political changes occurring in South Africa in the early 1990s when President FW de Klerk and Nelson Mandela of the African National Congress (ANC) negotiated an end to the system in 1993. Flying back to Johannesburg from visiting Eskom power plants, McRae noticed that large, dark, non-electrified areas surrounded major cities. That aerial perspective of the economic disparity helped him form a vision for change. Electrifying these areas made good economic sense for the utility executive. At great personal risk, he began a series of clandestine meetings with the ANC in 1987 to bring power to black townships under a policy of "Electricity for All." It took time for the ANC to embrace McRae's vision, but his later discussions with Mandela and his second-in-command, Cyril Ramaphosa, led to its acceptance. Its success brought electricity for the first time to black townships such as Soweto and Orange Farm, and rumour spread that Mandela wanted McRae to become his new government's minister of energy. However, McRae, who pushed for Koeberg to be among the first nuclear plants to host a peer review, greatly valued WANO and chose its presidency over a government position.[9]

McRae was a genial man with a ruddy complexion, bushy white eyebrows, a fine sense of humour, and a challenged – though enthusiastic – singing voice given to

Louis Armstrong songs. But it was his sincerity on nuclear safety, not his singing, that had made a favourable impression on WANO leaders, particularly Pate, after a speech he delivered at the Atlanta BGM. McRae volunteered Eskom's Koeberg plant, which he believed was isolated from the industry because of his government's racial policy, the plant's location and its operators' lack of contact with outside operating experience, to undergo one of the pilot peer reviews in 1992. He saw WANO as a way to end that isolation through exchanges and peer reviews. He made it a point to attend the exit briefing at a time when WANO members left such things to less-senior executives or plant managers. McRae found the peer review process extremely valuable and became an outspoken advocate for its permanent adoption as a WANO programme. Thereafter, Pate worked to convince him to become the next WANO president. At the Tokyo BGM in 1993, McRae was elected.[10]

McRae jumped into his new position and brought the same enthusiasm he had for singing to the goal of improving WANO's mission. "As an industry," he wrote in WANO's in-house magazine, *Inside WANO*, "our biggest long-term threat is complacency. We have seen so many examples of high-performing companies that have failed to recognise the signs of a deteriorating safety culture." WANO, he believed, could help those companies avoid that fate. He set out on a tour of all the WANO centres, speaking extensively with WANO staff. His first impressions, he reported to the Governing Board, reflected many of the general concerns of WANO members. Some of the issues were more directly linked to his own experience than to WANO's programmes, such as the public's tendency to link nuclear power with nuclear weapons, the ongoing problem of what to do with nuclear waste and the political problems of India and Pakistan, all issues then pertaining to discussions within the South African government. After the peer review, McRae clearly saw the relationship between the economics and cost effectiveness of nuclear power and how that equation was crucial to its future. Therefore, WANO's programmes were essential for all nuclear utilities. However, McRae quickly recognised that, although WANO's programmes were "commendable", problems remained, particularly "internal and external communication and the need to foster a stronger sense of commitment at the plant level". It was quite possible, McRae told the Governing Board, that "there were plants needing assistance which were unaware of their shortcomings."[11]

McRae's report was a prelude to the first Strategic Review Meeting of the Governors to review the effectiveness of the organisation and consider possible improvements to existing programmes and the need for new initiatives. Hosted by Atlanta Centre and INPO, the special session was conducted over two days in the spring of 1994 at Calloway Gardens, a resort complex outside of Columbus, Georgia, famous for its azaleas and not far from the late President Franklin D Roosevelt's retreat in Warm Springs. Carle thought the time was right to review WANO's operations. "In the past," Carle explained, "WANO had been able to build on the INPO model and although this had provided valuable guidance" the Governors now had to ask "what the future needs were and whether the original goals and objectives were still valid". Carle said that the purpose of the meeting was to give the Governors the opportunity to exchange ideas on a whole range of WANO issues with "a free and open discussion."[12]

Five years had passed since WANO's founding. McRae saw the anniversary as an opportunity to "reflect on how we have progressed and to decide what direction we should set for the future". Because of WANO, he noted, "there has been a more unified approach in maximising operational safety." Carle, too, reflected on WANO's short history. At the 1993 Tokyo BGM "there was a general feeling that WANO had established itself as a credible organisation, respected by the international community, having a clear role…in improving nuclear safety standards." But he cautioned the members that "complacency is our biggest enemy. The more successful we are at avoiding another accident, the greater the threat will become."[13]

Publicly, Carle urged WANO members not to relax their efforts, while realising privately that WANO was losing the intensity and commitment so critical to its founding. The involvement of senior utility executives was declining. As those respected leaders moved on, more junior individuals filled regional governing board posts. In addition, from an Atlanta Centre perspective, WANO programmes were not working well as they were implemented inconsistently among the regions. In American eyes, WANO had not achieved quite the international respect that Carle had declared.[14]

Carle noted that since its formation, WANO had the benefit of INPO experience. He asked Pate to review the work of INPO and highlight any lessons that might be relevant to WANO in the future. Pate noted one important difference: the prime objective of INPO was the promotion of excellence, whereas WANO's mission was to maximise safety. But in both organisations, he said, the "moving force was communication, comparison and emulation". Pate listed more than a dozen INPO principles, several directly relating to his audience. It was most important to secure the "interest and involvement of top management", he told the Governors, and that a "very special effort [is] required to overcome natural resistance to the acceptance and assimilation of operating experience at one plant into another plant's procedures and training." Also, he warned, INPO was "careful not to let promises get ahead of capability." Bill Cavanaugh, the President and COO of Carolina Power and Light Company based in Raleigh, North Carolina, and a member of the Atlanta Centre Governing Board, added that he believed that INPO's main success had been "to stimulate completely new thinking on the part of operators", something he hoped that WANO could translate to its members.[15]

As part of Carle's review of the value of WANO, the Governing Board asked Atlanta Centre to report on discussions regarding the future of INPO's International Participant Advisory Committee in the context of WANO's development. The purpose of the discussions was to determine if the time was ripe for a merger of INPO's international activities with WANO. Stan Anderson, the head of the International Participant Advisory Committee, polled his members. IPAC members preferred bilateral arrangements with INPO rather than the world connections of WANO. Utilities recognised INPO "for the high quality of its work, and, although WANO had achieved much at Kozloduy and central Europe, it commanded a much lower resource commitment and was not, therefore, seen in quite the same light as INPO". Additionally, WANO tended to lose its identity in some areas, such as exchange visits, where the technical teams consisted of INPO employees. After the first initiatives had been taken, Pate noted, "the visibility of WANO was gradually diminished." While IPAC had "the highest regard for WANO", he reported, IPAC programmes should stay in place and members would be "encouraged to strengthen WANO." Nonetheless, what made IPAC so attractive to WANO members was that it was a safe way to be involved with INPO and exchange ideas, but have no binding commitments. Any merger would have weakened WANO's Pilot Peer Review programme, which required a higher level of obligation than IPAC's programmes. Carle agreed that questions of WANO's visibility and value continued to challenge the organisation, and those issues remained on the Governing Board's agenda thereafter. But the idea of a merger with IPAC would not go forward.[16]

The IPAC evaluation review led to a spirited discussion on the value of WANO's programmes, revealing what the centres believed were useful. From the perspective of Paris Centre, Governor Ray Hall reported that WANO's Peer Review programme, workshops, experts meetings, and Twinning and Exchanges programme were of the highest value. The Event Reporting programme was "useful" to some utilities, but most others failed to provide event reports, share relevant experience or exploit the capabilities of Nuclear Network®, leaving much room for improvement in this area. The Performance Indicator programme varied widely, being embraced by some plants and ignored by others. Paris rated the Good Practices programme "poor" as operators did not respond to suggestions on paper in the same way they did to practical demonstration. In general, he said, the commitment to WANO on the part of top executives had declined and needed to be reinvigorated. The question, McRae asked, was how could WANO persuade its members to become more fully involved? The Governors from the other centres agreed that Hall's analysis on the usefulness of the programmes and commitment of the members reflected their own views.[17]

The lack of commitment to WANO began with the attitude of the utilities in the regional centres. The Governing Board thought that the staff at the regional centres were not as complementary or mutually supportive as they would like. Moreover, staffing in some centres was inadequate in terms of numbers and skills to meet WANO's requirements. As a case in point, Moscow Centre, even with the extensive assistance provided by Atlanta Centre, lacked the necessary funding to meet the goals of WANO's programmes, including the Peer Review programme, a situation readily admitted by its Governors. Language issues and the costs of translators and interpreters added to its woes. Tokyo Centre also admitted to "problems which only it could solve" but declined to identify them further. Tokyo's major sticking points, however, remained language issues and its opposition to the WANO Peer Review programme, even though WANO did make adjustments in the way peer reviews were conducted and reported in order to ease the centre's cultural sensitivities. But the major failing, all Governors agreed, was the tenuous interrelationship between the regional boards and their respective utility members. Regional board members, the Governing Board suggested, should make a constant effort "to maintain and improve such links by all means of contact, including plant visits by Governors".[18]

Communications was another issue that vexed the Governors. All agreed that more should be done to publicise WANO's achievements both within and outside the organisation but, because of regional differences, a communications plan to cover all of WANO was difficult – if not impossible – to devise and implement. Internally, the London Coordinating Centre produced a quarterly newsletter, *Inside WANO*, initiated in November 1993, and regional centres distributed brochures, collected video footage for later use and drafted scripts for question-and-answer briefings. Recognition of WANO came from an unexpected source when Clarke, the Director of the Coordinating Centre, was made an Officer of the Order of the British Empire (OBE), in part for his contributions to nuclear safety worldwide.[19]

Nonetheless, the regional centres often emphasised locally developed programmes. Tokyo Centre took great pride in developing "WANO Caravan", which consisted of two or three staff members travelling to power plants to inform operators of WANO's activities, thereby increasing their interest in participating in WANO and its programmes. The initiative demonstrated, the centre director reported, that "operators were still not well informed on the aims and objectives of WANO". Moscow Centre said it was necessary to promote "the feeling of the 'need' for WANO" and stressed the importance of frequent meetings between plant managers and the regional governors. Externally, the Governing Board urged senior WANO officials to speak at key conferences, brief other nuclear industry organisations on WANO activities and target newspapers for additional coverage of WANO. Carle thought that external communication remained a major challenge for WANO, particularly after the third BGM in Paris in 1995, when, he said, "the impact of the media in Europe was virtually nil."[20]

Carle also raised the issue of whether or not WANO's role in central and Eastern Europe through the Users' Group and Coordinating Committee should continue.

Moscow Centre Governors – Oleg Saraev, Director of the Beloyarsk nuclear power plant, and Kozma Kouzmanov, the Director of the troubled Kozloduy plant – strongly supported the work of WANO in Russia and the former Soviet Bloc nations in sanctioning the Users' Group and establishing the Coordinating Committee. Kouzmanov praised WANO's "successful" intervention at his power plant where WANO had been "instrumental in removing the electricity shortage in Bulgaria." WANO, he said, "had created the basis for effective management and development of a whole new approach to operation, maintenance, quality assurance and safety". But Saraev added that Marshall's project "should not necessarily be [WANO's] main focus", a position gaining broader currency throughout the association. Breaking with the Users' Group was not a simple process. As Marshall would later remind the Governing Board, the Russians had been crucial to the formation of WANO. In return, WANO agreed not to criticise Russian plants in public and to provide what assistance it could. He implied that dropping the Users' Group would ignore that important history. The consensus of the board was that WANO should continue its "tacit support" for its central and Eastern European activities, but that its assistance should wind down over the next year or so.[21]

In addition, Carle believed that a discussion of potential new programme initiatives should be included in a wide-ranging review of WANO activities. At the top of the board's list was the possibility of WANO providing rapid assistance to a plant in the event of a serious accident. Although such assistance was not in WANO's Mission Statement, several Governors supported the idea, as did McRae, who recognised this need for any utility whose plant was geographically isolated. What he envisioned was WANO compiling a list of experts ready to provide assistance and advice on short notice. The Atlanta and Paris Centres reported that they already had drawn up such a list, but it had never been used. Tokyo Centre explained that its Japanese utilities had their own response plan, and Moscow was discussing a system of self-help to cover not only expertise but also spare parts. While the Governors supported the idea in principle, they left any level of implementation to each centre, citing language and international response times as reasons for WANO not to assume leadership. With each centre proceeding independently, the Governing Board concluded that assistance would be a utility-to-utility matter and that WANO would act "solely as a facilitator" for such arrangements and not dispatch teams to troubled plants. The fact that the Governing Board often deferred to the regional centres on aspects of the core programmes, as well as on adopting new ideas, reflected WANO's governing structure, where real power resided in the four regional centres. It was an arrangement that increasingly frustrated those who sought a more active and effective WANO, a higher level of participation by senior management, and organisation-wide remedies to achieve that end.[22]

The activities of WANO's Tokyo Centre were especially weakened by political problems, which included seemingly insurmountable language and cultural differences, festering wounds from World War II and sharp differences between the governments of India and Pakistan and between Taiwan and mainland China. Tokyo Centre, according to one WANO official, struggled to implement WANO programmes because Asian cultures were embarrassed by failures and reluctant to share problems, preferring to solve those matters privately. Many governors believed that the failure of India and Pakistan to sign the Treaty on the Non-Proliferation of Nuclear Weapons (commonly known as the Non-Proliferation Treaty, or NPT) and a determination by the Japanese utilities, led by TEPCO, to define WANO's core programmes on their centre's own terms had sapped WANO's effectiveness. The absence of staff from the Asian sub-continent placed more pressure on the Japanese utilities to fill those positions, an additional cost they were hesitant to assume, leading to underfunding and understaffing at the centre. Tokyo Centre complained, with some justification, that "there are so many differences among our members [that] it is difficult to apply a single policy or working method to all members equally." The Canadian government withheld assistance to India's CANDU reactors, and other countries followed the same NPT-based policy, to the detriment of WANO's work in Pakistan. Engineers from India and Taiwan could not travel to Pakistan, and Pakistani peer review team members could not enter Canada. As a result, meetings for all members could only be held in Japan or Korea. The Governors were also troubled that India had not fully informed WANO about a turbine fire and shutdown at the Narora nuclear plant in the northern state of Uttar Pradesh, releasing only an internal report to Tokyo Centre rather than to the WANO Governing Board, thereby endangering WANO's credibility and causing "considerable disquiet among the governors".[23]

Although divided, many of the governors had also become "disquieted" about the Users' Group, headed by Lord Marshall. Formed in June 1993 by operators of Soviet-designed reactors, the Users' Group sought to work together to upgrade the safety of those operations. While it was functioning well in their opinion and Lord Marshall had achieved a "remarkable result in difficult circumstances", the group "still had some way to go to achieve its objectives". Specifically, the board worried about the extended timetable and the implications that held for additional funding, particularly a Special Assistance fund established to underwrite Marshall's activities that had gone over budget. Marshall's penchant for overspending had been more or less tolerated during his chairmanship, but many on the Governing Board had grown impatient with nagging delays and cost overruns from the Users' Group. A majority of governors wanted the project accelerated and, if possible, to find funding for Marshall outside of WANO. Moscow Centre continued to support the Users 'Group, arguing that it had finally identified "some eight to ten common projects to be completed". To others, that was not much of an achievement after more than a year. They stressed that WANO should support neither an open-ended time frame nor an open chequebook. In a compromise, the Governing Board agreed to fund Lord

Marshall for the next six months; after 1995, the board wanted alternative funding agencies to underwrite this work and there would be no formal links between WANO and the Users' Group.[24]

By the end of Carle's first term as Chairman, the Governing Board had completed its review of WANO's programmes. The verdict was mixed – Atlanta Centre was performing above expectations, others below what the governors expected. Nevertheless, Carle was upbeat, though with a caveat, reporting to WANO's members that overall the board was "satisfied that the programmes were on course and producing good results". However, he warned, "we have to be on our guard against complacency that could rob us of all that has been achieved." He noted that although there had been a reduction in the number and importance of abnormal events, "some mistakes are still being repeated and we must do more to improve our analysis and the communication of results." He called on WANO members to do more in sharing the results of exchange visits, to make more effective use of performance indicators, and to renew efforts to identify areas of good practices. Although Carle noted publicly that these challenges were a "normal part of the development of a still young organisation" the undercurrent among the Governing Board was that these issues had been slow to change. There was broad agreement that support from senior utility executives had declined in some regions and that top management had sent lower-level plant managers in their stead. WANO needed to be valued at the highest levels, and senior utility management needed to be convinced of the benefits of participating in the organisation's activities.[25]

That Carle demanded better results did not indicate that WANO's programmes were failing. In fact, the opposite was true for several of them. The Operator to Operator Exchange programme, initiated at the Inaugural Meeting in Moscow with the challenging goal that each Moscow Centre nuclear plant would arrange a reciprocal visit with a plant from another region by the spring of 1991, had succeeded beyond expectations. Those initial visits had given the organisation a strong foundation for continued exchanges. By the end of 1994, more than 250 exchange visits had taken place. The programme had also matured. The focus of the exchanges moved from general information to specific operational and managerial issues identified in the earlier exchanges such as maintenance, training, radiological protection, plant organisation, chemistry, quality assurance and emergency planning. Workshops, seminars and expert meetings, which provided the opportunity for power plant personnel to focus on specific subjects, grew and matured like the exchange programme, playing an increasingly important role in sharing operator information and advancing WANO's goal of maximising nuclear safety and plant reliability.[26]

WANO's Performance Indicator (PI) programme was a keystone to the goal of plant reliability. Most utilities kept internal records of their plants' performance. But the WANO programme permitted them to benchmark against utilities from all over the world. WANO's performance indicators "let us know where we stand and how we compare" with other members, according to one site manager. "WANO's PI's are the cornerstones of our own 'what gets measured, gets done' approach, which is the most powerful management tool we all know." By mid-1994, 95% of WANO's members reported on seven of the ten indicators. Nuclear plant performance continued to improve, with unit capability increasing from 72% to 80% between 1990–1994. There were similar gains in the unplanned capability loss factor, in lower collective radiation exposure, and in the number of unplanned automatic scrams. Utilities with good performance indicators translated into better managed, more reliable, and safer nuclear plants. Nevertheless, WANO members agreed that while the programme had been "extremely effective in providing a clear picture of trends in nuclear plant performance", there was still room for improvement, especially in the definition of the indicators and the reporting process. The Governing Board created a task force to determine why members of Paris Centre failed to meet the expected level of reporting performance indicators in line with WANO goals. The task force learned that there had been problems in communication and that WANO needed to market the programme better. As a result of the investigation, there were "encouraging signs there would be significant improvements in the participation of members". Atlanta Centre, which co-ordinated the programme for WANO, proposed changes to improve the sharing of performance indicator data and increase its usefulness for WANO members. By 1996, Atlanta Centre had produced a CD-ROM containing a complete historical archive of performance indicator data, as well as plant-specific information for better comparison – the latter a result of a relaxation of confidentiality rules regarding performance indicators by the WANO Governing Board to make the programme more effective. A year later, members could download the reporting information from the WANO website.[27]

Of all the WANO programmes, peer reviews had been the most difficult to incorporate at the Moscow meeting and, perhaps, the most successful since. Though some delegates – primarily from Eastern Europe and Asia – rejected peer reviews as part of the original WANO core in 1989, Bill Lee, Pate and others at INPO who recognised the value of peer reviews had lobbied hard for WANO members to begin them in some form. The compromise, worked out in 1991, was the Pilot Peer Review programme. The success of the pilot reviews led to the establishment of the Peer Review programme in 1993 as one of the basic WANO programmes. Lee and others pushed successfully for review teams to visit a certain number of plants each year, providing the visits did not interfere with the review by the International Atomic Energy Agency's (IAEA) Operational Safety Review Team (OSART). This ensured that, although WANO and IAEA plant inspections were different, they were complementary and not competing.[28]

WANO tailored its reviews to meet regional issues. As a result, peer reviews were done in accordance with specific WANO performance objectives and criteria that had been approved by the regional directors. Initially, most of the expertise for conducting the reviews came from INPO, which made a concerted effort to train nuclear professionals from other nations in the peer review process, including representatives from each region in each review. Carle pushed for integration of the review teams, insisting that each consist of experts from all regions. By creating these mixed teams, Carle sought to break down the cultural uneasiness among the four regional centres. Often led by Atlanta Centre staff, the team consisted of representatives from up to 10 different countries and the four regional centres, and would spend two weeks at each plant. Atlanta Centre also assisted Moscow Centre in its peer reviews, enabling Moscow to conduct its own peer reviews in the future. By 1995 the teams were functioning smoothly and able to conduct reviews and follow-up visits without a large American presence.[29]

In an industry once known for its lack of communication between plants, international nuclear operators eventually came to value the peer reviews. "I was a little nervous about hosting a peer review at the Tomari nuclear power plant," admitted Tatsuo Kondo, the plant's manager. "After it was all over, we discovered it was a very positive experience. The peer review not only helped us to recognise our strengths, but also to focus on areas for improvement." Pavel Ipatov, Director of the Balakovo nuclear plant about 900km southeast of Moscow, spoke for many members who embraced the value of peer reviews: "Contacts made with peers during the review process play a significant role in enhancing nuclear safety and plant reliability by actively using the world's accumulated operational experience." Nevertheless, some members complained that there was too much emphasis on social events such as elaborate and expensive team dinners, rather than the intense work experienced reviewers thought necessary.[30]

Kondo's comments aside, Tokyo Centre remained an unwilling partner in the Peer Review programme, insisting that it, rather than WANO, define what consisted of a peer review. "Few Tokyo Centre members volunteer for a peer review and they cannot be forced to host one," declared Masateru Mori, a WANO Governor. His members did not see the value of a peer review, and the high cost of translation and interpreters also operated against the visits. If WANO did not modify the peer review process, Tokyo Centre members would not volunteer. The sticking point for the Japanese was that peer reviews identified problems rather than providing solutions to problems, as a programme focusing on technical services – which the Japanese preferred – would. In the end, the Governing Board reached a compromise. While Carle would not modify the terms of a peer review, WANO would accept the terminology of technical service visits from Tokyo Centre members. Peer reviews run out of Tokyo Centre remained quite different from those in the other regions.[31]

Peer reviews also received a cold shoulder from German plants, which were reluctant to participate because they were hesitant to divulge information that might be used by the powerful "Green" lobby against nuclear power in spite of WANO's confidentiality policy regarding the distribution of peer review reports. When the Governing Board approved peer reviews in adopting a five-year plan in 1996, it noted that "there will be some differences in the implementation of programmes between regions." Nonetheless, Carle emphasised, it was "essential that all members follow principles and requirements" of WANO's programmes. By then, peer reviews interlinked all of WANO's programmes, demonstrating that peer reviews now represented "the key WANO programme".[32]

PROGRAMMES OF **WANO**

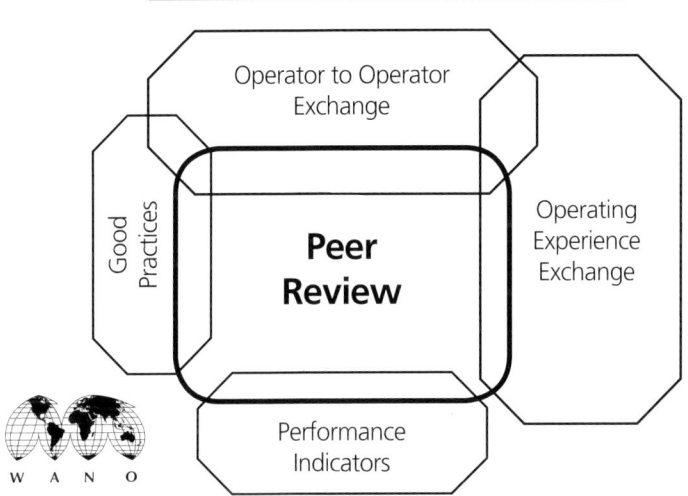

Diagram developed by WANO in 1995 to demonstrate the interlinking of its five core programmes and the central role of peer reviews

Two WANO programmes, however, did not match the success of peer reviews and operator exchanges. After some initial success following the Moscow meeting, involvement in the Operating Experience programme, which had experienced much success among INPO members, fell off as fewer WANO members submitted event reports. The programme always had its detractors, particularly utilities in regions that believed the sharing of event reports over the Nuclear Network® electronic messaging system was more harmful to their reputation than the value of learning from common experience. In addition, not all the Regional Centres fully developed the ability to analyse and communicate event trends to their members. While WANO noted that the drop in reporting operating events might be a function of the improvement in plant performance indicators, the staff warned that "the open sharing and learning from our experience is essential for the success of the nuclear industry."[33]

The second programme to disappoint WANO's staff was the Good Practices programme. Regional centres were to identify and develop good practices depending on the needs in each region and the resources available in each centre. While the number of identified good practices increased steadily from 1990–1994, from five to 95, just 28 were identified for WANO generally. "In order for this programme to achieve its maximum effectiveness," the Coordinating Centre believed, "identified good practices must meet the needs of our members by providing ways to resolve problems and issues they face on a daily basis." But much needed to occur. "As we obtain information about strengths and good practices from participation in other WANO programmes, this programme will continue to mature." By 1995, lists of good practices were actively collected but, because they were only in English, rarely consulted.[34]

Election to two consecutive terms as chairman was not established by the WANO constitution. However, the precedent set by the members in electing Marshall to a second term was carried over with Carle, who continued in his position unopposed. Indeed, it took the better part of two years for a chairman to visit most of the power stations, listen to the concerns and comments of utility executives and plant operators, learn to work through the various strengths and weaknesses of the regional centres, judge WANO's effectiveness and begin to develop a plan for building on WANO's mission of achieving higher levels of safety and reliability at its members' plants. Meeting in a closed session in Seoul, Korea, in November 1994, the Governing Board unanimously recommended that Carle be elected to a second two-year term and that Eric Pozdyshev, the president of Rosenergoatom, be elected the third WANO President at the 1995 BGM held in Paris. Carle, who was set to retire from EDF in 1995, planned to devote more of his time to WANO activities. Pozdyshev, who trained as a physicist at Leningrad State University in the late 1950s, began his early career working with plutonium production reactors at the Soviet weapons complex near the city of Krasnoyarsk on the Yenisey River in Siberia. A decade later he moved back to the Leningrad nuclear power plant to work on the construction and commissioning of the first two RBMK reactors, which were based on the design of the graphite production reactors. After the Chernobyl accident, Pozdyshev was appointed Director of the Chernobyl nuclear power plant with orders to restart Units 1 and 2 and to mitigate the consequences of the accident.[35]

****

Many WANO members agreed that the 1995 Paris BGM marked a new phase in the history of their association – WANO's start-up phase had ended. Over its initial six years of operation, WANO had proven itself despite internal challenges and considerable political and financial upheaval in the countries of many of its members. Marshall had safely navigated the association over the shoals of cultural and political differences. The Moscow Centre/Paris Centre partnership had demonstrated that WANO could expand its core mission and respond to a special need cooperatively and without recrimination in order to improve the safety of the world's most troubled reactors. Peer reviews, once opposed so strongly that they were not included in the initial WANO mission, had become a central focus and the organisation's most successful programme. Each year more WANO members became trained by INPO as peer review evaluators. In addition, participation in the Performance Indicator programme was steadily increasing and the results were measurably positive. During WANO's early years, Pozdyshev suggested, "we relied on and benefitted from the experience of organisations like INPO for our ideas and methodology. Now we are relying more on our own experiences and ideas and defining a WANO perspective and way based on those experiences and ideas." In other words, the new President of WANO believed that it had "achieved the vision set forth in that inaugural meeting in

Moscow in 1989. We are now a mature organisation positioned to strive for excellence in our WANO programme operations."[36]

Pozdyshev was only partially correct in his assessment. Certainly the start-up years had ended, but excellence remained elusive. WANO still faced many of the same challenges the Governors had identified earlier – communication issues, financial constraints, a lack of top-level support from many utilities and reluctance on the part of some members to view WANO as a "credible and cost-effective source of help in improving operating performance". Programmatic success had been sporadic. The pressures on the industry caused by the Chernobyl disaster had faded into the past, and the urgency for safety that the accident had engendered diminished. In addition, utilities everywhere, but especially those with the emerging governments of Russia and Eastern Europe, were under great pressure to economise, making it difficult for plant operators to see the benefit of an organisation that asked them to share experiences in order to stop something from happening.[37]

Carle recognised these changes and pushed to restore the sense of urgency in safety matters among all members as part of a larger goal for WANO, reminding them that the nuclear power industry was "as weak as the weakest among us". He envisioned an association, not unlike INPO, "with widely known credibility and consistently valued results so that station managers automatically think of WANO when facing new challenges or seeking to improve performance". He stressed that "the biggest dividends accrue[d] from peer reviews," both to individual stations that volunteered to host them and to the rest of the worldwide nuclear community. The larger challenge for WANO members, he said, was "to make similar strides" in WANO's other programmes. To maximise the operating safety and reliability of nuclear power stations throughout the world and achieve standards of excellence, Carle said, there needed to be better communication among members and "a substantial increase in the use of WANO programmes, more requests for peer reviews, increased use of operating experience, more operator to operator exchanges, improving performance indicators and more use of good practices." His second term would be devoted to strengthening these core programmes.[38]

In order to achieve his goal of a stronger, more engaged WANO, Carle believed that he needed to reinvigorate member commitment to WANO's goals rather than the regional variations favoured by some centres. That emphasis became the theme of the 1995 Paris BGM. Carle wanted the CEOs to reaffirm their commitment to WANO and return from the meeting with instructions for their plant managers to participate fully in WANO's programmes. Based on discussions at the BGM, the Governing Board agreed that the basic aims and objectives of WANO were being met. However, in reviewing the meeting, the board inquired as to the effectiveness of the regional centres. Carle suggested that Vince J Madden, who had replaced Clarke in July 1995 as the Director of the Coordinating Centre, investigate the implications of making the regional centres twice as effective, including increasing their resources. Madden's CV was ideal for the job. A physicist by training, Madden was a 31-year veteran of the UK nuclear power industry, involved with design, commissioning and operational aspects of nuclear plants as well as working on nuclear safety, training and performance issues. Prior to joining WANO, he was responsible for Nuclear Electric's HQ Operational Standards unit, a post that covered operational experience feedback, human factors, performance indicators, the UK's peer review programme and the interface to WANO and INPO for all four UK nuclear utilities. In addition, he had international experience in working with the IAEA, UNIPEDE and the G24 Technical Working Group.[39]

Carle followed up his theme of strengthening WANO's goals among its members at the following BGM. Since its founding, WANO had rotated the location of its

BGMs among the four regions – Moscow, Atlanta, Tokyo and Paris – between 1989 and 1995. The 1997 meeting was the responsibility of Moscow Centre. In 1995, newly elected WANO President Eric Pozdyshev said he wanted to hold the 1997 meeting in Moscow. However, the Governing Board said no and asked the Russians to move it to one of the other Moscow Centre member countries. Both Ukraine and the Czech Republic offered to host the meeting, the former in Kiev, the latter in Brno. Unable to come to a decision, Anatoly Kontsevoy, the Director of Moscow Centre, asked the London Coordinating Centre for assistance. Andrew Clarke agreed to visit both cities to determine if they had adequate facilities to host the conference. In Kiev there was only one hotel that could accommodate the number of attendees. "We stayed at the hotel and I could see things weren't promising," Clarke recalled. "While they could provide you with a hotel bed, they didn't serve breakfast. You had to go somewhere else in town for that." Clarke spoke to the hotel manager. "How many rooms do you have?" he asked. "I can't tell you," she replied, "that's a state secret." Clarke tried another approach. "If I were coming here and wanted a meeting for 400 people in two years' time, could you accommodate them?" She burst out laughing. "If you were coming next week with 10 people I couldn't tell you whether I could accommodate them or not," she said. "Every time I book somebody in, somebody from the government comes and chucks them out and says 'I'm going to stay here'." There was no venue in Kiev that proved satisfactory.[40]

With Kiev out of the picture, Clarke flew to Vienna, rented a car and drove to Brno, not far from the Austrian-Czech border. Brno, Clarke found, could accommodate the conference but the city was small and difficult to get to, a four-to-five-hour drive from Prague, which was much more convenient. Carle insisted that WANO meet in Prague, a far more appealing city with a worldwide reputation for its architecture and beauty. In addition, Prague had an efficient conference facility at the Hilton-Atrium Hotel, located between the business district and the picturesque Old Town and considered one of the best convention venues in Eastern Europe. The Governing Board quickly approved Prague for the next BGM. "The Czechs were wonderfully organised," Clarke remembered, "and everything went like clockwork."[41]

The fourth BGM in Prague in May 1997 marked the end to Carle's four years as WANO's Chairman. While he was enormously proud of what WANO had accomplished since its founding, his extensive travels for the association since his retirement from EDF revealed a number of issues that made him uneasy for WANO's future. First of all, two pioneering giants in the founding of WANO had died the preceding year. In January, Lord Marshall, who had shaped the organisation and held it together in its formative years, passed away after a long battle with cancer. "Having presided [at] the creation of our Association," Carle said, "he modelled it and made it what it is today. He used to say that the existence of WANO was a miracle in itself; he was the maker of that miracle." The following summer Bill Lee, WANO's first president, died suddenly of a heart attack while visiting New York City. Lee, Carle stated, should be remembered for "his enthusiasm and his practical sense which he knew how to communicate with liveliness and persuasion. Let's try to communicate our convictions to all sceptics around us with the same strength." With the character and commitment of both men in mind, Carle presented his chairman's report to the General Assembly. "I have often looked in my mind for what they would have said if they had been among us at this biennial meeting." Neither man, Carle hoped, would have disowned his report, in which he emphasised the organisation's strengths and reminded the delegates of the serious work yet to be done.[42]

Chernobyl continued to lurk over the industry. However, in the eight years of WANO's history, Carle noted, there had been no nuclear accidents. "I don't believe that this is merely the result of some mathematical probability." WANO, its members and their employees had made nuclear power operations safer. Nevertheless, Carle explained, "this is not about preserving a technological option, but about excluding humanly unacceptable events" such as Chernobyl. WANO had to do better "because we will never do enough to erase that original sin." Moreover, WANO now possessed the tools to accomplish its mission. "If these tools were largely inspired by those of INPO," Carle explained, "they had to be adapted to the international context and now they are WANO's own tools." If WANO used these tools effectively, Carle believed, it would have a bright future.[43]

The first important tool in WANO's arsenal was the event report system to collect and analyse reports on incidents that could communicate one plant's experiences to other operators who would ask themselves: "Can this happen to me, and what should I do to avoid it?" Carle saw event reports as a necessary step leading to a real safety culture. "Nobody knows what safety culture means if he has not gone through this process," Carle exclaimed. "I am always afraid that those who never experience any incidents may in fact be short-sighted and do not see the abyss they walk around until they fall into it." Yet outside the US, members were reluctant to report incidents to others in spite of the fact that "an incident at any one plant affects us all." By failing to share event reports, members denied their colleagues the opportunity to learn and potentially prevent a similar incident that might be of greater consequence. But the level of dialogue that Carle thought necessary took place "only if it is initiated and encouraged by the regional centres" which, he believed, had not done all they could to achieve good communication between plants and report on events. "We have this shared responsibility," he reminded the members, demanding that they renew their commitment to WANO and to event reporting.[44]

Other tools included programmes such as visits, twinnings and seminars; performance indicators; and good practices. The first had initially proved effective, especially the twinning arrangements and workshops and seminars, which had been for the WANO members from Eastern Europe the decisive programme in opening their plants to the outside world. But by 1997, interest in these programmes had waned. Carle believed that members needed to communicate better with the regional centres to find topics of more interest and value. Performance indicators were constantly evolving. The programme offered plants the opportunity to make comparisons and to open discussions for improvement. Good practices, on the other hand, were rarely used, largely because the programme consisted more of specific plant good practices than WANO good practices. Carle hoped to change this equation and bring the programme more in line with the needs of the entire membership.[45]

The Peer Review programme, however, had become the most effective of WANO's tools, Carle believed. The WANO peer review, he reminded members, was not an evaluation or an inspection, "but work in common, an exchange completely in line with the spirit of our association, the central point of WANO activities". He ticked off the reasons to oppose peer reviews: they were expensive, operators would not accept being judged, there was a language barrier, reports could be used against the operator and "we already have our own internal review system." Nevertheless, Carle explained, the programme was "so essential" to WANO that these reasons for not instituting the programme were flawed. No internal review could "replace the experience brought by external operators. Even the best among us still has something to learn." There was a language barrier, he said, but it was easily overcome in situ among people of the same profession "with a lot of bad English and a little interpretation". At the end of a peer review, he noted, "there are only friends and a splendid mutual understanding, whatever the so-called 'cultures' are". While peer reviews had costs, those costs were small in comparison with total operating costs, and the return was high. And finally, Carle promised that all WANO peer review reports would remain confidential, although at the time that concept was being tested in Canadian courts. Carle was appalled that the courts would order the public disclosure of sensitive information. "There is for me something inadmissible in this [Canadian ruling]," assuring the members that confidentiality would remain WANO's policy.[46]

The main difficulty with the Peer Review programme, Carle explained, was the lack of availability and training of peer review team members. For years INPO had provided the bulk of team members and had trained WANO regional centre members in the procedure. While WANO was grateful for this assistance, Carle said, the time had come for all the centres to build their own trained teams. "A WANO Peer Review is not an INPO evaluation. It has to take into account the international aspect and the diversity of cultures involved, to define its own style, which may be different from one region to another. However, we do not want four WANO Peer Review programmes. All centres will have to continue to work together, pooling their experiences to maintain adequate consistency."[47]

During his tenure as Chairman and years later, Carle worried about gaps in WANO's internal and external communications activities. Along with making WANO's programmes more effective, Carle called on WANO members to increase communications among members and to publicise its activities with the external world. "I often notice during my visits to plants that WANO is hardly known outside a small circle of people." He saw the newsletter, *Inside WANO*, and the caravan organised by Tokyo Centre as "excellent initiatives" but WANO needed to do more. In addition, he fretted that "very few referred to WANO at the time of the 10th anniversary of the Chernobyl accident." He hoped that in the future WANO would "be a place of reference for all that relates to the state of operation of our power plants." To accomplish these goals, Carle called on WANO members to double their funding for the regional centres. Doing so "would allow them to be much more efficient and to increase dramatically the contribution of WANO to nuclear safety. Why don't we sell WANO with all the resources publicity and marketing can offer?"

he asked.[48]

One example of WANO's major contributions to nuclear safety had been its active presence in all the Eastern European plants since the establishment of the Paris Centre/Moscow Centre Committee under Lord Marshall. WANO continued to occupy a central role as an organisation with no national interests but considerable industry interest in promoting the safety of the troubled Soviet-designed reactors. In a time of suspicion and mistrust stemming from the Cold War, a long tradition of Soviet secrecy, Chernobyl, and considerable financial difficulties, WANO had been non-judgmental, a position not lost on the Russians and Eastern Europeans. From this experience, WANO had demonstrated a framework for achieving a climate of trust and collaboration. In Carle's longer view of this accomplishment, he saw WANO spearheading an "interconnected nuclear system across the whole of Europe." He also envisioned that the regional centres would increasingly "devote special efforts" to assist countries when they faced economic difficulties. "In a world that becomes smaller," he said, "our contemporaries will not be able to accept that national frontiers can justify differences in the standards of protection of their citizens, in respect of their health and environment." Carle called on a "new WANO," a revitalisation of its programmes and a remobilisation of its members to work toward excellence. "If we do not do it, WANO will only be, in a few years, an exhausted bureaucracy and a beautiful memory."[49]

In the spring and summer of 1996, Carle and the Coordinating Centre staff, working with the Governing Board, drew up WANO's Long-Term Plan for 1997–2000. The Plan was, in fact, a strategy for improvement. At the top of the list was "a special concern over the rotation of WANO staff, including the management of the centres." The Plan also called for improved selection and training of staff for the regional centres. Other aspects were consistently on WANO's list of areas for improvement, such as more and better member involvement in its programmes and improvement of internal and external communications, particularly with the public, media and key people in other organisations. The Governing Board assigned oversight for the Plan's implementation to the director of the Coordinating Centre, a move that raised the hackles of the regional centre directors. Carle was adamant, however, insisting that the regional centres must support the Coordinating Centre director, and the Governing Board unanimously approved the Plan.[50]

Carle also saw a need for two regional centres, Moscow and Tokyo, "to be more cooperative". In a heated encounter in Tokyo, he urged an influential former member of the centre's governing board to replace the centre's director, whom Carle considered obstructive to WANO's programmes. Though this change was very difficult for the Japanese to swallow, Carle finally prevailed. He wanted all centre directors to be "team players". The new Director of Moscow Centre, Farit Toukhvetov, a former Station Director at the Bilibino nuclear power plant, agreed with the team approach. He told *Inside WANO* that he would work with his centre's members to increase their participation in WANO's activities, placing "a special emphasis on new WANO initiatives to address existing problems and to anticipate future challenges". If the top-down approach – from the chairman and governing board to the regional governing boards and centre directors – had not accomplished WANO's mission as quickly as Carle had hoped, perhaps a shove from the broader membership could get the regional governing boards back on a single track to move WANO ahead. Carle wanted to ensure that all WANO members would buy into its strategic plans for the future, especially peer reviews – which he believed needed more volunteers in some regions – and sharing of operating experience. This might be accomplished by listening to as many members as possible about the strengths and weaknesses of WANO – its valued programmes as well as its less useful ones and how they might be improved.[51]

To accomplish this, Carle and the WANO Governing Board called for a full review of the organisation, a "self-examination" led by Raymond W Hall of Paris Centre and Robert C Franklin of Atlanta Centre, "two elders we all respect." Ray Hall had a long career in nuclear energy, initially with the CEGB and later as Chief Executive of Nuclear Electric and of Magnox Electric Limited. For many years Hall had served as a member of the Paris Centre Governing Board and, after 1993, as its chair. Bob Franklin, a former Canadian railroad executive, had become the head of Ontario Hydro in 1986 and led that utility's entrance into nuclear power. In addition, he had chaired Atlanta Centre's first governing board. Both men were highly regarded in the industry; both would be retiring soon and would be able to devote much of their time travelling and speaking to WANO members throughout the world. Their task, as Franklin explained to the Atlanta Centre board, was to review all WANO activities by asking opinions of as many personnel as possible regarding the organisation's strengths and weaknesses, focusing on WANO's programmes, the performance and resources of regional centres, and WANO's structure. During the summer and early autumn of 1997, the team would investigate members' views on WANO's programmes, with a goal of completing their report to the Governing Board by the end of the year. Their report would be Carle's legacy.[52]

****

WANO under Rémy Carle had matured considerably. From an uncertain organisation largely held together by the force of Lord Marshall's personality, WANO had become accepted within the international atomic energy industry as a force for nuclear safety. Marshall had traded the centres' autonomy in return for WANO's stability. Intractable problems between the regional centres and the WANO Governing Board were often postponed until some later date for the sake of general harmony. Carle believed that the time had come to emphasise WANO's goals and core programmes and, while "accepting their differences", bring the regional centres more in line with the association's guidelines rather than their own. To drum up support, Carle, usually accompanied by Clarke and occasionally Pate, travelled extensively to meet with utility executives and plant operators. By including more executives in his discussions, Carle hoped to strengthen regional support for WANO's programmes. These contacts forged strong relationships with members, building on and expanding Marshall's work in Eastern Europe and Russia and in the Far East. "He was a fine statesman for WANO," Pate later commented regarding Carle's visits to WANO members' plants, "and highly dedicated to the [WANO] mission." He had witnessed the growth of the Peer Review programme and new efforts to improve communication among members with an expanded Nuclear Network® and the publication of *Inside WANO*. The safety and reliability of nuclear plants had improved, and Carle believed that WANO's programmes had contributed to this. But, he warned, "behind this progress lies a decisive improvement in safety culture, but which remains threatened wherever the financial situation is precarious. We should make nuclear energy a model technology and aim for excellence. There is a margin of progress which we should and can fulfill whatever country we are in." Implementation of the Long-Term Plan and the Franklin-Hall Report, he expected, would provide the roadmap by which WANO could succeed.[53]

Chapter Four

SECURING THE MANTLE
OF NUCLEAR SAFETY

The election of Dr Zack T Pate to the chairmanship of WANO in Prague in May 1997 marked another stage in the progress of the organisation. From the time of its founding, WANO members had been reluctant to place Americans in positions of authority. Americans, including Bill Lee and Pate himself, who were present at the creation of WANO in both Paris and Moscow, agreed that for an international organisation to succeed, Americans could not lead it. That Pate was chosen to head WANO eight years later at the fourth Biennial General Meeting was a testament to the leadership of his predecessors and the comfort level with – and respect for – Americans that members had acquired during the operation of the international association since its formation. For almost a decade, INPO, with Pate as President and CEO, had provided large financial, technical and personnel contributions without accompanying demands. The one programme pushed by Americans – peer reviews – had proven to be one of the most successful. The fear of WANO becoming an international INPO had so far failed to materialise. For many WANO members, Americans were no longer seen as an all-powerful force in dictating nuclear safety principles but as a valued ally within WANO in establishing worldwide standards for nuclear operations.

The Governing Board was well acquainted with the person it was electing. For years Pate had been a distinguished, highly regarded figure within the international nuclear power community. A forceful President and CEO of INPO for more than a decade, he had vigorously supported the creation of WANO and had served as an active and vocal adviser to the WANO Governing Board and the board of the Atlanta Centre.

Pate's nuclear safety credentials were flawless. A 1958 graduate of the US Naval Academy, he volunteered for submarine duty. Pate later commanded a nuclear submarine and served as an aide to the legendary Admiral Hyman Rickover. Pate was steeped in Rickover's demand to achieve the highest nuclear safety standards, a lesson reinforced when the sister ship of a submarine he was serving on, USS *Thresher*, was tragically lost during deep diving trials in 1964. After two tours at sea, Pate accepted a delayed Burke Scholarship earned at the Naval Academy to study nuclear engineering at the Massachusetts Institute of Technology (MIT), where he worked with noted professors Michael J Driscoll, acknowledged as an outstanding teacher in the field, and Norman C Rasmussen, an expert on risk and reactor safety and the developer of probabilistic risk-assessment techniques that would lead to new approaches in evaluating nuclear reactor safety. Pate's PhD dissertation, *Severe Reactivity Excursions in Fast Reactors*, made him keenly aware of the behaviour and potential danger of nuclear reactors and the critical need to operate them safely and efficiently.[1]

In the weeks after the Three Mile Island accident in March 1979, Pate, then a member of Rickover's staff, outlined plans for a "Commercial Reactor Plant Inspection Organisation", loosely based on the US Navy's Operational Reactor Safeguard Examinations Board (ORSE), which conducted inspections of the nuclear fleet and held the ships and crew to demanding safety standards. Failure to meet those standards could result in shutting down the reactor. Pate envisioned an inspection organisation established and funded by the nuclear industry but able to operate independently of the industry and its regulator, the Nuclear Regulatory Commission.

This new organisation would conduct annual inspections, rate the plants and have the authority to stop operations at a plant if necessary. Once Pate learned that the Kemeny Commission was recommending the creation of a utility organisation along the lines he was proposing, his career path became clear to him.[2]

Pate had no desire to remain in the navy. At age 43, after 22 years of service and having risen to the rank of Captain, he decided to retire. In the spring of 1980, Pate left military service to join the newly created Institute of Nuclear Power Operations (INPO), an electric utility industry organisation based in Atlanta not far from his boyhood home of Leesburg in southwest Georgia. INPO held great promise, he believed, and its mission of emphasising nuclear safety echoed Pate's own views. Moreover, Pate admired INPO's President, Eugene "Dennis" Wilkinson, a retired Admiral who had been Rickover's pick to command the first nuclear submarine, the USS *Nautilus*, as well as the first US nuclear surface ship, the USS *Long Beach*. Both men embraced the strong nuclear safety culture as dictated by Rickover and believed it could be applied beneficially to the civilian power industry.[3]

There was one difference between the two men in their approach to how industry might adopt that culture, however. Wilkinson would work with industry executives to win them over to supporting INPO in part because nuclear safety made good economic sense. Pate, on the other hand, was less patient and more willing to challenge the utility CEOs to adopt and achieve INPO's safety standards. A strong supporter of performance standards, peer reviews and plant evaluations, Pate pushed INPO utilities hard during his years as CEO of INPO, including shutting down one plant and forcing another utility's board of directors to remove the company's chairman and president and bring in a new management team focused on improving safety. Pate was one year from stepping down from his INPO post when he became Chairman of WANO. His views on the importance of nuclear safety and the central role and accountability of utility executives in achieving safe operations had not changed.[4]

WANO members had seen their new chairman at every Governing Board meeting and BGM, where his passion in promoting worldwide nuclear safety had cast a keen and steady eye over WANO's development. Confident and experienced from his time with Rickover and at INPO, Pate could be both a tough taskmaster in advocating change at WANO and a careful diplomat in dealing with WANO's diverse membership. He was well aware that if opposition to American leadership at WANO had dissipated, it had not disappeared. Pate was every bit as physically imposing as his unshakable commitment to nuclear safety. Tall and athletically trim, he possessed a rich bass voice that embodied authority. When Pate entered a room, he captured it – the remarkably sonorous voice, the shining shaved head standing above the crowd and his whip-sharp knowledge of nuclear safety all demanded attention.

Just as he had been an early advocate for an industry response to the Three Mile Island accident, he had been in the vanguard to establish an international organisation soon after the details of the Chernobyl accident became known. The cause was just, but the timing, for Pate, was not. On the 25 April 1986, he and his wife Bettye arrived in Bermuda to celebrate their 25th wedding anniversary. It was their first vacation in more than five years. They had been there less than 24 hours when the first news of Chernobyl arrived. Soon Pate was on the phone to INPO to discuss what could be done. "The end result," he later recalled, was that "I didn't get a lot of credit from Bettye for a vacation."[5]

Pate and others at INPO were uncertain how to approach the disaster. In the middle of the Cold War, he had visited the Soviet Union during his days as a submarine commander, looking through a periscope at the lights of Murmansk and secretly monitoring shipping traffic moving in and out of the harbour. But this was hardly the

CV with which to open discussions with the Russians. In addition, Americans in the mid-1980s questioned the sincerity of *glasnost*, a policy more publicised than practised at the time. Nevertheless, Pate and others at INPO were determined to find a way to communicate with the Soviets and create an international approach to nuclear safety. INPO could serve as a model, but the Americans agreed that an international INPO would not work.[6]

When Pate finally set foot in the Soviet Union, it was to attend the WANO Inaugural Meeting in Moscow in 1989. To the Soviets and others he was not just another delegate. In the months prior to this first gathering of the world's nuclear utilities, Pate directed part of INPO's staff to draft the organisational guidelines and functional programmes for the new association. As a result, there was a prominent stamp of INPO's experience on the WANO Charter. In addition, Pate and his wife were travelling with Lord and Lady Marshall, whom they had joined in London a few days before the meeting. Their first stop was Leningrad and a visit to the Leningrad nuclear power plant some 70km west of the city on the shore of the Gulf of Finland. The plant, once managed by Nikolai Lukonin, had four Chernobyl-type RBMK reactors. Pate was impressed with the attention and deference Marshall and he received in Leningrad.[7]

But the enormous gap in tourist accommodations between the Soviets, who were new at hosting international conferences, and the modern hostelries in the West with which Marshall and Pate were familiar, became evident when the two couples checked into the convention centre hotel in Moscow. "The Russians were treating Marshall as a visiting dignitary, and they lumped Bettye and me into that [category] because they had heard of INPO," Pate remembered. "We thought we were getting the royal treatment on the coat-tails of the Marshalls, who were assigned a suite. So they take our baggage and escort us to our room, give us the key, and leave. Actually there were two rooms, with a door between. In one room there are two double beds. In the other room two sofas, two coffee tables, and several chairs. Marshall and I soon realised that this is how the hotel had created a suite." For the next half-hour, Marshall, a Knight of the British Empire, Pate, the CEO of INPO, and their wives moved furniture and reconfigured the rooms. "We didn't know who to talk to," Pate explained, "so we just fixed it." Soviet egalitarianism and Western ingenuity had won the day.[8]

The tour of Chernobyl following the Moscow meeting had a great impact on Pate. Based in a hotel in Kiev, the WANO group took a hydrofoil up the Dnieper River to the vicinity of the plant and boarded buses for a tour of the ghost town of Pripyat, a short distance from the Chernobyl plant. While they were not allowed into the plant, Pate took pictures of the city. "Every tree was dead. You could see curtains open and towels hanging on the balcony. People had just abandoned it. Trucks were spraying water on the streets to keep radioactive dust down." It was a stark, jolting reminder of the extent of the accident. Rickover, INPO, Moscow and Chernobyl all served as milestones in Pate's history and in his preparation for becoming Chairman of WANO.[9]

****

When Pate assumed the Chairmanship in mid-1997, WANO became headed for the first time by two North Americans. Canadian Allan Kupcis, the CEO of Ontario Hydro, was elected the new WANO President. A 24-year veteran of the utility, Kupcis was born in Riga, Latvia, and emigrated to Canada when he was eight. He earned a PhD from the University of Toronto and did postdoctoral studies at Oxford before joining Hydro's research division in 1973. Kupcis's field was material science, investigating the metallurgy of the reactors and developing non-destructive internal tubing and reactor vessel testing and inspecting tools. Later Kupcis took a few courses on management at MIT and rose through the management ranks as a corporate planner. When the politics of the Ontario provincial government, the utility's owner, shifted,

Kupcis became the Acting President in 1992. A year later he was elected President and CEO. The two men had worked together at INPO through its International Participant Advisory Committee, and Pate strongly supported Kupcis's candidacy as WANO President. Pate had a set of goals he wanted the WANO Governing Board to achieve during his Chairmanship, and Kupcis – a strong supporter of peer reviews – was an important ally in that quest.[10]

Kupcis had become a powerful advocate of peer reviews through his own experience at Ontario Hydro. In his presidential speech at the 1997 Prague BGM, Kupcis stunned his audience with a *mea culpa* accounting of the slippage in his company's nuclear performance. "For the past few years," he said, "the management performance at our nuclear units has been sliding and past programmes to address this have been slow in stopping the slide. Ontario Hydro is a graphic illustration that even the best can fall from grace through the inattention of senior management to the nuclear business. Quite frankly, we became complacent. We believed our own press clippings." He warned that "when management of safety suffers, performance suffers and business suffers, too. Our costs have been escalating and unplanned shutdowns of our units cost us dearly in lost income last year." What had turned the company around, Kupcis explained, were WANO "peer reviews that revealed the extent of the slide in our nuclear performance". The first review was conducted at the Bruce nuclear generating station on the eastern shore of Lake Huron in 1992. Bruce was the first non-US plant to sign up for a WANO peer review, which became a "turning point in understanding how insular Ontario Hydro had become in its safety culture." Kupcis revealed that "we have learned from those peer reviews…and are conducting a comprehensive assessment of our people, our plants and every aspect of process performance at each one of our nuclear stations." WANO reviews, he concluded, were instrumental in breaking down Ontario Hydro's "insularity" and improving "the safety of our operations". Kupcis's message was clear: WANO peer reviews were essential to the continuous improvement of nuclear facilities. It was a message he would continue to deliver during his presidency.[11]

Pate had listened carefully to Rémy Carle's concerns about the condition of WANO and the need for an internal review of the organisation. Carle argued that it was time to evaluate how WANO was meeting its mission after eight years, and in early 1997 WANO began the review process. Pate fully backed the idea of an internal review. As WANO Chairman, Pate molded his agenda with that in mind. He sought to improve member participation and cooperation in the activities of the regional centres, to "make sure there was a good chairman and a good director in each region." He also hoped to put Moscow Centre on a more solid financial footing and build it into an effective safety centre for former Soviet Union plants. In addition, he encouraged the centre to occupy improved office space and enlarge its staff. A fourth item on Pate's agenda was to achieve universal peer reviews, to get all the members to accept obligatory peer reviews, "not just volunteer once a decade". Ideally, WANO plants would have a peer review every two years, following the INPO model, but he knew this would be difficult to achieve. Nevertheless, Pate wanted to deal swiftly and effectively with weak-performing plants, and peer reviews were an important step in managing safety concerns. Finally, he wanted to be certain that the Franklin-Hall Report, which WANO members approved at the Prague BGM and sent to the Governing Board by the end of 1997, got a vigorous and positive response from each region. "Al and I were fully in sync on how to approach the Board with these proposals."[12]

The timing to hold an internal review of WANO was ideal. There were changes in the leadership at all of the centres, with new regional governing board chairmen taking over in Tokyo (Morgan Tsai from Taiwan), Moscow (Aleš John from Czechoslovakia) and Paris (Willy De Roovere from Belgium). Ryosuke Tsutsumi became the new Director of Tokyo Centre, and Dr Farit Toukhvetov replaced Anatoly Kontsevoy, who had served as Director of Moscow Centre since 1991. René Vella remained the Director of Paris Centre. Atlanta Centre retained its top leadership: William Cavanaugh III remained as the Chair of the regional governing board and Sig Berg continued as the centre's Director. WANO was at a point where former leaders could express their concerns about how WANO functioned, and the new group of officials, Pate and Kupcis hoped, could adopt proposed changes as their own and assure their implementation. If there was one concern that Pate had with the personnel changes, it was that the new WANO Governing Board members, as well as those coming on to the regional governing boards, did not enjoy the same top-level executive positions as those they replaced. And Pate had always stressed the importance of top executive involvement.[13]

With the new makeup of the regional governing boards, Pate and the WANO Governing Board wanted to put as much pressure on the incoming members as possible to ensure that their responsibilities to WANO would not be slighted. In an effort to reinforce the obligations of WANO members to the association and to strengthen the regional boards, the WANO Governing Board urged regional governors to meet with senior utility executives to seek increased commitment to WANO. With WANO members in 31 countries, Pate realised that it would be physically impossible for him to visit all the sites with Kupcis and Vince Madden, who replaced Clarke as Director of the London Coordinating Centre. Drawing on his predecessor's creation of a new position, Special Ambassador, for Marshall, Pate named five Special Ambassadors to assist the regional governors, plus Rémy Carle to serve as a Special Ambassador for the WANO Governing Board. Ray Hall and Bob Franklin, who were just finishing their internal review of WANO, would cover the Paris and Atlanta Centres respectively. Ryo Ikegame was named the ambassador to Tokyo Centre and Evgeniy Ignatenko for Moscow. The appointments were an ideal method, the Governing Board agreed, for furthering the implementation of the recommendations of the forthcoming internal review conducted by Hall and Franklin. The aim of the ambassador programme was to meet face-to-face with as many WANO members in all the regions as possible to build relationships and support for WANO. "I was emulating Carle's idea in creating these positions," Pate later said. "It was a good thing. It helped."[14]

****

Toward the end of Carle's chairmanship, he, Pate, Madden and the WANO Board held a number of discussions about reviewing the progress of the organisation. The genesis of the idea for a review was a feeling on the part of several WANO executives that they needed to hear more from the members. They had seen some improvement in the regions with operating experience and with peer reviews, but the pace of improvement, the executives believed, was too slow and "wasn't what we wanted it to be". The Governing Board decided on an internal review of the organisation that had two major objectives. The first was to give members a voice in what WANO was doing and where it was headed. The second goal would emerge from those conversations – to analyse how WANO was meeting its mission after eight years in operation. All agreed that Ray Hall, who was British, and Bob Franklin, a Canadian, should conduct the review, in part because both were native English speakers and English was the official language of WANO.[15]

With a six-month deadline to complete their review, Hall and Franklin fanned out over the world, interviewing nearly 200 senior utility executives, plant managers, regional board members and staff in all regions and all regional centres. It quickly became evident that many members saw the same strengths – and weaknesses – in WANO's operations and programmes. At a Governing Board meeting in Vancouver, British Columbia, in September, the two men provided a preliminary report of the general impressions of the members' perceptions of WANO. Most agreed that

WANO's founding principles were still valued and that the organisation had "done most things right" and achieved much. Nevertheless, "WANO's existing mandate," the team reported, "is not yet totally fulfilled." Perhaps not surprisingly, there was a dichotomy of views expressed by senior managers and plant managers, primarily that there was greater support at the corporate level, "but much less at [the] plant level". Often plant managers complained that corporate financial constraints and cost-cutting placed additional pressures on WANO participation. The Operating Experience Information Exchange programme, Franklin warned, was "regarded [as] good in theory, but [had] little use in practice" and needed urgent attention from the Governing Board. While plants "are deluged with reports", most of the event reports were not of high quality and lacked credibility. As a result, many plant managers "valued only INPO reports". Finally, there was a widespread belief that WANO's momentum was slowing down and that the organisation "will not take full advantage of the work already done".[16]

At the Governing Board meeting in Atlanta in November 1997, the two men presented a detailed account of their findings. Nothing had changed. Responses, they admitted, had ranged from "enthusiastic to indifferent". Nearly all the individuals interviewed agreed that WANO was an important and valuable addition to the nuclear community and should continue as a non-governmental, voluntary, decentralised association with "a degree of regional autonomy", which translated as "respecting language and cultural differences", meaning the regions could implement programmes as each saw best. The idea of a central authority was not popular. Members viewed WANO as a "partnership of peers". WANO, most members said, should focus on safety and reliability. A minority wanted more emphasis on cost-reduction programmes. A fear that American culture would dominate WANO was never far from many respondents' minds, as they saw the organisation as being modelled on INPO and dependent on its resources. "Problems of differing language, culture and affluence permeate all [WANO] activities," and WANO had not "managed these differences well". Nevertheless, the team concluded, "radical changes are not necessary" and "more a mid-course correction and a rededication to WANO" could address the association's shortcomings.[17]

Pate thanked Franklin and Hall for their work, adding that it "fully met" Carle's vision to carry out a peer review of WANO. Their review had "given the Board an excellent opportunity to take WANO forward". Nonetheless, the Governing Board could not take WANO forward alone. Not only would the Governing Board need to find a consensus on responding to the report, it would also need the full cooperation of all the regional centre directors and governors. Their ability to work with the WANO Governing Board to achieve the improvements recommended by Hall and Franklin would determine whether or not the existing administrative structure of WANO would continue.[18]

Pate invited comments from the Governors regarding the Franklin-Hall Report. They generally agreed that there was a need for an early response in some of the areas "in order to demonstrate that the Board is serious about the review". Beyond that there was "considerable debate" centering on WANO's core programmes. The Governing Board decided to realign the Technical Exchange programme, bringing technical support, operator exchange visits, good practices, performance indicators and WANO Nuclear Network® under one umbrella, subsequently renamed the Technical Support and Exchange programme. Staff also presented the Governing Board with a proposal for improving the Operating Experience programme, which took significant lumps in the internal review, but no action was taken at the meeting. The Peer Review programme and Professional and Technical Development programmes would remain largely unchanged. In response to member feedback, WANO redesigned *Inside WANO* into a full-colour quarterly newsletter and published it in English, Russian and Japanese, with translations into French, German and Spanish. The publication was part of a major WANO effort to make communications with members more relevant and readable.[19]

****

Aside from the Franklin-Hall Report, the Governors dealt with two other issues, one old, one new. The inability of Moscow Centre to meet its financial obligations had created a dilemma for WANO for several years as the Russian Federation and newly independent countries of the former Soviet Union struggled to put their chaotic economies in order. From the beginning, WANO's Governors had recognised the importance of the participation of the former Eastern Bloc nations, and the wealthier centres made up much of the annual shortfall of contributions to the operations of the Coordinating Centre. To remedy the situation, the Governors established a surplus funding strategy that allowed the director of the Coordinating Centre to seek emergency funding from other centres when reserves fell to a certain level. For three years running, from 1996 to 1998, members from Ukraine failed to pay their membership fees to Moscow Centre. In addition, some Russian plants, including Leningrad, also did not pay, and the Governing Board struggled to find a way out of the problem without embarrassing the plants in arrears and adding to the financial burdens of other members, suggesting that Rosenergoatom make up the shortfall. Rosenergoatom had pledged to pay current Russian plant fees to Moscow Centre, Ignatenko reported, but this did not include old debts. By the end of 1997, Moscow Centre had run out of cash and minimised activities, and the government had closed all Rosenergoatom and Russian plant bank accounts. A separate legal process had to be taken in order "to allow payment of wages to plant personnel", Ignatenko said. The situation in Ukraine was "even worse than Russia", he warned, adding that predictions about when these debts might be repaid were "difficult to make".

Although the financial situation in Eastern Europe seemed an intractable problem, "helping Moscow Centre", Franklin and Hall noted, "is regarded as crucial, with wide, but reluctant, support."[20]

The second issue concerned the status of WANO's newly elected President, Allan Kupcis. Forced from his position with Ontario Hydro by the provincial government, Kupcis no longer held a position in a nuclear utility that qualified him to serve as WANO's president, a situation not unlike Marshall's when he served as WANO's chairman and lost his position at the Central Electricity Generating Board. Kupcis offered to resign, but if asked, he would continue to serve. In his acceptance speech as WANO President, Kupcis had used the example of what Ontario Hydro had learned from WANO's voluntary peer review of the Bruce nuclear generating station in 1992 in "understanding how insular the company had become in its safety culture issues". As a company, Ontario Hydro had "fallen down with respect to our safety culture. We believed our own headlines of how great we were." Kupcis had challenged the WANO members in Prague "to think seriously about making sure that they take part in WANO activities and not get isolated or arrogant in terms of their own operations". It was an approach to WANO members that Pate and others on the Governing Board wanted to retain. They prevailed upon the Canadian utilities to provide financial support for Kupcis, and a potential crisis was averted. The Governing Board also proposed an amendment to WANO's Articles of Association to insure that "if the president is unable to complete the normal two-year term between Biennial General Meetings, a WANO regional governing board may propose candidate(s), and a successor may be elected by the WANO governing board, to serve on an interim basis until the next Biennial General Meeting."[21]

****

While the Franklin-Hall Report and finances occupied much of the Governing Board's discussion, the repercussions of a WANO peer review of the Chernobyl plant in the early summer of 1997 demonstrated the strengths and weaknesses of the organisation. Around the time of the 10th anniversary of the accident, the WANO Governing Board became concerned with the safety of the one operating reactor at Chernobyl, Unit 3. By 1997, Unit 4 was a sarcophagus; Unit 2, which had been badly burned in a fire in 1991, was not operating; and Unit 1 had been long closed. The WANO Governing Board determined that it was time to conduct a peer review at the plant. The leader of the peer review team was Michael Hayden, an American who had been with the IAEA for many years and the head of a number of OSART teams. He had earned an excellent reputation at IAEA for his work, and WANO picked him to lead the Chernobyl review, "thinking he would have additional credibility in Ukraine."[22]

The peer review would not be cause for a 10th anniversary celebration. Chernobyl was the 51st peer review conducted by WANO and, according to one team member, "the problems were of *much* greater concern that those found in any previous review." The international review team, which spent three weeks at Chernobyl, found numerous areas for improvement for Unit 3. The situation was so bad that Hayden was brought in to brief the WANO Governing Board at its meeting at the Le Méridien Hotel in London in July. Hayden explained that the team had found eight strengths and 14 areas for improvement (AFIs) during the review and summarised the more significant ones for the Governing Board. Working conditions were difficult. The operators were bussed in and back home every day, past the burned-out plant, which had not been cleaned up. Several modifications to Unit 3 that were designed to prevent another accident had not been started or completed. One particularly worrisome area was the failure of the plant to correct fuel enrichment and "the positive coefficient of reactivity at certain control rod configurations", both major contributors to the initiation and seriousness of the 1986 accident at Unit 4. In addition, the team found "an abnormally high number (74)" of reactor coolant steam leaks at the top head of the reactor, with some of the steam plumes "as tall as a person". Typically, one might observe three or four steam leaks through the head structure of an RBMK. Moreover, the leakage was so bad that the humidity prevented some key instrumentation from working properly. In addition, there were serious corrosion and maintenance issues, ineffective fire protection measures and "widespread lack of respect for radiation protection procedures and practices". The morale of managers and staff was low, Hayden reported, "because of pay shortages, lack of finances to correct equipment problems and uncertainties over the future of the station". The circumstances greatly "distressed" the WANO Governing Board. Pate later wrote that "the situation at Chernobyl was the worst I had ever seen, by a considerable margin."[23]

At the end of the discussion, Pate informed the Governing Board that Chernobyl "would probably be the most significant and worrisome report of all WANO reviews conducted so far". The plant's condition required a vigorous response, Pate believed. The WANO Governing Board acted as never before. The Governors unanimously agreed to send a letter from the WANO chairman to the owner/operator of Chernobyl, listing the key information from the peer review report and "stating that conditions are not conducive to continued safe operation" of the plant. In addition, due to the severity of the case, the chairman also wrote to the top government officials in Ukraine urging them to correct the deficiencies, and the WANO Governing Board sent a letter to the IAEA advising it of the peer review results. The Governing Board also decided to inform the governments of those board members who were part of the G7/8 nations – the US, the UK, Germany, Canada, France and Japan – urging those governments to lean on Ukraine to fix the plant or close it. So important was the issue that the Governing Board, going "beyond its normal confidentiality practices", sent a letter to the European Commission regarding the peer review. Finally, the Governing Board asked all the regional chairmen to inform their members in writing of the situation at Chernobyl and enclose Pate's letter to the plant's owner/operator. The various actions, Pate later recalled, "resulted in an unprecedented (in my experience) blitz of correspondence, dialogue, and travel!"[24]

The WANO chairman's letter to the deputy minister of energy in Ukraine was concise. Unit 3 was vulnerable to a serious fire such as had occurred in Unit 2 in 1991, and the "safety culture problems that contributed to (or led to) the accident in 1986 have not been addressed. The safety culture shortfalls, in combination with the degraded equipment and lack of completion of safety modifications, result in an unacceptably low margin of nuclear safety. It is our conclusion that urgent action is necessary to resolve and correct this unacceptable situation."[25]

Pate made it clear to the Ukrainian government official that WANO peer reviews were voluntary and that the Chernobyl plant manager volunteered for the peer review. "He and his staff were fully cooperative with the WANO team during their visit" and had requested assistance from WANO subsequent to the review. "The plant manager and his staff must not be made 'scape goats' or victims of this situation," Pate stressed. "They recognise they need help, and many of the problems are beyond their control."[26]

Although the Moscow Centre governors on the WANO Governing Board supported the strong steps taken to remedy the problem at Chernobyl, the resulting pressure from Western governments put the Moscow Centre Governing Board in a delicate and very unwelcome position. "It was a really, really tough deal at the time," Kupcis remembered, when the Russians demanded to know why WANO, contrary to its Charter, had involved a government "with issues that concerned [only] WANO". The regional governing board "felt that we were treating Chernobyl unfairly and creating another disaster in the eyes of the world. It was seen as WANO stepping out of bounds by our Russian members and, I think, by some others around the world, too. They thought 'what kind of role was WANO taking now?'" Farit Toukhvetov, the new Director of Moscow Centre, complained that WANO should have limited the release of the information about Chernobyl's problems, arguing that it was "more important to help the plant". He urged the WANO Governing Board "to think about what tools can be used to help a plant in the future," to create "a set of actions which demonstrate the ability to help a plant in difficulty." His pleas, however, were late. Apart from the WANO initiatives under Marshall and the Paris/Moscow team, the West had sent millions of dollars to aid Chernobyl and other Soviet-designed plants, with limited effect. Some observers, including Pate, believed that WANO's tough actions following the peer review accelerated the post-Chernobyl modifications of other RBMK reactors and were a key factor that led to the final closure of Chernobyl Unit 3 in 2000.[27]

****

The fifth BGM was held in Victoria, British Columbia, in September 1999, 10 years after the WANO Inaugural Meeting in Moscow, and boasted the largest attendance of any WANO meeting to that time – more than 350 delegates representing every country with operational nuclear power plants. Given that Victoria, sited at the end of Vancouver Island at the eastern edge of the Pacific Rim, was not a major city or transportation hub, the attendance figures were all the more remarkable. Held at the magnificent Empress Hotel, a classic railroad hotel in the old European tradition, the meeting was exceptional by nearly any standard. WANO had taken over the entire hotel. Then, when delegates arrived and learned that the hotel staff had gone on strike, the hotel's management and WANO event staff coordinated to ensure the conference ran smoothly. They succeeded. "Most of the people who attended…couldn't even tell that the staff were on strike," Pate recalled. "I think what we gave up was the turn-down of the beds." The delegates dined next door at the Royal British Columbia Museum, drinking in the province's rich history and culture. Later, attendees boarded boats for offshore whale watching and toured Butchart Gardens, a world-famous floral showplace. The hotel's location overlooking Victoria harbour, the planned activities and the WANO programme contributed to a fascinating introduction to Canada and a memorable BGM. The election of the new WANO President, Soo-byung Choi, President of the Korea Electric Power Company, was a complement to WANO's presence on the Pacific Rim.[28]

Pate wanted to keep the meeting upbeat, to examine WANO's progress and re-emphasise the importance of high-level participation. The meeting's keynote speech, given by former US Senator Sam Nunn, commended WANO on its mission. The organisation's stress on safety, cooperation across national boundaries and "transparency among peers – must become the example" for the world, Nunn declared. "The ability of the world's nuclear operators to exchange information and achieve higher levels of nuclear safety is unparalleled in any other industrial sector." Pate told the delegates that the record level of attendance was "a clear signal that the senior executives of the world's nuclear utilities continue their commitment to cooperation on nuclear safety." The attendance figures were equally a result of the success of Pate, the Governing Board, the regional directors, regional governing boards and Special Ambassadors to convince top executives to attend and rededicate their commitment to WANO. An earlier proposal to the Governing Board that utility officials dedicate themselves to WANO in a "re-signing" ceremony reenacting the Inaugural Meeting in Moscow 10 years before, had been rejected. How deep that commitment was, however, would remain an unanswered question.[29]

After 10 years, WANO officers thought, on balance, that their organisation was flourishing, although it was evident that significant issues remained and needed to be addressed. Yet in spite of large linguistic and cultural differences among members, the convulsive geopolitical changes that had taken place since its organisational meeting and the dramatic shifts in leadership at nuclear utilities in the wake of those political transformations, WANO endured. The organisation's cohesive single focus on its mission of the safe operation of nuclear plants, Kupcis believed, was the key to WANO's longevity and continued success as a voluntary organisation. Marshall and Carle had each taken WANO one step further. The proof was in the statistics. In 1989, WANO members represented 144 plants; in 1999, they represented 202 stations with 443 commercial reactors in 31 countries. In 1992, the first year of peer reviews, WANO conducted four; 26 were planned for 1999 alone. WANO reached a milestone that year with the 100th peer review at the Krško nuclear power plant in Slovenia. Information sharing, performance indicators, exchange visits and event reporting had all improved. Yet after 10 years, the memories and imperatives of Three Mile Island and Chernobyl had begun to fade, and WANO officials warned members not to become complacent.[30]

The task ahead was not easy. During his first term, Pate himself had met with more than 30 utility executives in all four regions. The landscape for electric utilities, Pate learned, was "undergoing tumultuous change". Deregulation and restructuring in the industry around the globe with associated competition and cost pressures worried many WANO members. "One of the nuclear industry's most important challenges – and therefore one of WANO's most important challenges – is to ensure that economic pressures do not cause nuclear operators to take shortcuts with safety. Despite the intensity of the competition that is developing, and the distractions that rapid changes in the environment are bringing about, nuclear safety must remain our highest priority."[31]

Although WANO celebrated what had been accomplished during its first decade –

"A Decade of Progress," WANO proclaimed – the Governing Board wrestled with plans for the association's future. As WANO prepared to enter the 21st century, it looked beyond all the Y2K predictions of disaster to develop a long-range plan that would respond to the Franklin-Hall Report and strengthen its programmes to meet the demands of a shifting industry landscape.[32]

In response to member feedback during the internal review, WANO developed a more extensive programme to publish reports that would "help identify important safety issues at the plants and provide information for investigation of the vulnerabilities of the stations to these issues". Significant Event Reports (SERs) and Significant Operating Experience Reports (SOERs) were written for complex events. The SERs consisted of an event summary and a listing of "significant aspects and underlying causes" to help plant managers verify that adequate processes were in place or that corrective actions had been taken to prevent a similar event from occurring at their station. SOERs focused on an event or related events occurring in a significant problem area "important to nuclear safety or plant reliability", such as the 1997 WANO peer review of Chernobyl that revealed that the plant again required prompt attention. The SOERs also contained recommendations to remedy such problems. Starting six months after the publication of each SOER, WANO planned to "assess the implementation of the SOERs recommendations in peer reviews at the stations".[33]

The SERs and SOERs were part of a larger effort by WANO officials to re-emphasise and redeliver the core mission and core programmes of the association. Just as the internal review sought to boost the active involvement of senior utility executives whose commitment to WANO had faded since its founding, so WANO's 10th anniversary gave the organisation an opportunity to re-examine and re-emphasise its key goals. Included among these were peer reviews; the collection and increased use of operating experience information; higher-level usage of the electronic communications through the WANO website and WANO Nuclear Network®; to upgrade, refine, and bolster the use of WANO's performance indicators and improve their usefulness and effectiveness; and to improve internal and external communications to "encourage participation in WANO programmes and to position WANO externally so that it is perceived as a credible organisation" by the end of 1999.[34]

****

The predicted catastrophic storm of Y2K software malfunctions turned out to be a minor passing shower for WANO's computers as the calendar turned over to the year 2000. Nevertheless, the organisation continued to wrestle with some of the same issues identified in the internal review – WANO's mission, its public profile, non-participative members, deregulation in the electricity industry and potential new safety initiatives. At Pate's urging, the Governing Board agreed to hold a special strategy session with Governing Board members and regional centre directors in November in Cape Town, South Africa, to weigh the pros and cons of each item. Pate explained that the Governing Board would be "simply examining these topics and that this was not a forecast of major change". As a backdrop to the strategy session, Pate stated that WANO's "organational structure [had] proven to be sound and effective", its policies and programmes had been widely accepted and member plants had steadily improved over the past decade. But referring to a nagging challenge, Pate said WANO needed to achieve a greater level of participation by those members whose participation was minimal. He also sought to improve resources for some regional centres as well as the quality of WANO programmes. The Cape Town meeting, Pate said, would provide WANO's Governing Board with the "opportunity to examine the need for changes to programmes, membership and policies in WANO's second decade to meet the future needs of the worldwide nuclear industry."[35]

During the first two days of the Cape Town meeting, regional directors aired concerns regarding their specific challenges. As each director reported, Pate provided his summary and offered suggestions for each centre. William Kindley, director of Atlanta Centre, commented on the need to focus the centre's programmes on individual plants in response to the problems revealed in peer reviews. Although Atlanta Centre had provided a significant amount of support to WANO, Pate suggested that the centre should give employees from other centres more opportunities "to experience US nuclear plant peer reviews firsthand". If peer review team members were to attain a certain level of ability, "the door must be opened to all regions."[36]

None of the regional reports held surprises, though the chairman's remarks revealed the delicate nature of issuing criticism leavened with counsel. Farit Toukhvetov explained that Moscow Centre's greatest challenge was involving more plant personnel in staffing the centre's programmes. To work around the lack of resources, Rosenergoatom had pushed programme participation down to each plant. Pate said that the plan was a good means to compensate for the lack of resources and urged the centre to encourage the Ukrainian nuclear power plants to become more involved with WANO. John Moares, Director of Paris Centre, noted that his greatest challenge was also staffing, not for lack of resources but from staff turnover and difficulty in securing replacements. While commending Moares, Pate asked him to push Électricité de France to take a greater role in WANO activities. "EDF provides support to WANO," Pate observed, "but is not making use of WANO as a resource."[37]

Pate reserved his harshest criticism for Tokyo Centre. The Japanese nuclear safety programme had experienced a major setback with a recent accident at the Tokai-mura reprocessing plant that resulted in two deaths. Although the reprocessing plant was not a WANO member and was not under WANO scrutiny, it did reveal an underlying issue of safety culture in Japan. One underlying problem was that Japanese nuclear operators were reluctant to hold peer reviews. To Pate's dismay, they also shunned any WANO technical support. He said that the Japanese nuclear utilities relied on INPO for technical advice and therefore did "not put sufficient resources into WANO. This has resulted in Tokyo Centre not facing up to the need for technical support missions," to the detriment of other members in the region that were "prevented from attaining maximum benefit from WANO". In sum, the major issue that concerned the chairman and the London Coordinating Centre was that the WANO regional centre directors did not work together to move WANO forward at "a common pace".[38]

The strategy discussions demonstrated the distance between the Governing Board and regional governing boards on most issues. Pate had said he expected no policy agreement from the discussions, and he was correct. Caution ruled. Even though there was considerable support for expanding membership to include vendors and contractors in recognition of their increasing role in the day-to-day operation of WANO members' plants, there was general consensus that "a very deliberate pilot project" should be considered before taking any formal action. On the issue of raising WANO's public profile, there was no consensus; however, there was one suggestion that WANO "first raise its profile with its own members". The group did agree that current WANO programmes were the best way to deal with the changes occurring in the industry through deregulation, competition and consolidation – and, with the potential dangers stemming from cost-cutting, a new owner's lack of appreciation of nuclear energy's special safety requirements, declining plant performance or other detrimental market forces. The issue of limited or no participation fostered no agreement and was settled as it had always been – discussion would continue. Finally, an attempt to form a WANO "nuclear safety advisory council" split the Governing Board and some of the regional directors, who saw this proposal as having the potential to weaken the regional centres. One participant reiterated that the four regional governing boards should serve as advisory groups to the main Governing

Board, arguing that a new advisory group was not necessary and would add another layer of bureaucracy to WANO's organisational structure. However, WANO's Governing Board had often acted without input from the regional boards, which were often caught by surprise and outnumbered in the decision-making process.[39]

As his second term neared completion, Pate pushed for a large turnout to the sixth BGM, scheduled to take place in Seoul, South Korea, in September 2001. Attendance numbers were critical, he told the Governing Board. If Tokyo Centre members sensed a weakening of support from other regions through low attendance, he warned, "it will discourage them from participating in the future." In addition, Pate issued a second challenge to the WANO governors. He wanted to be able to say in his speech at the BGM that "every plant has had or has volunteered for and scheduled a peer review," and he asked each director and governor to work toward that vision.[40]

By mid-2001 WANO had largely completed what one Coordinating Centre director termed its "developmental phase". If its first years under the chairmanship of Lord Marshall constituted a "start-up" phase that created an administrative structure to deal with the regional centres and enabled the organisation to be up and running, the years under Carle and Pate's chairmanships witnessed the full definition and development of WANO's four core programmes – Peer Review, Operating Experience, Technical Support and Exchange, and Professional and Technical Development. Everything seemed in place to advance WANO to the next level after the Seoul BGM.[41]

Then two events intervened.

The first event was internal to WANO. Choosing a successor to Pate proved exceedingly difficult. At an executive session in the Lake District of England in the summer of 2001, the Governing Board could not settle on an acceptable candidate.

The Governing Board interviewed two potential nominees. The first aspirant failed to convince the Board that he would do the job as they defined it; the second threatened to pull his region out of WANO if he were not selected. The Governing Board, seeking a gracious way out of a thorny and politically delicate situation, chose neither interviewee and decided to extend the search. The Board asked Pate to stay on for a fifth year, giving ample time to find a suitable replacement. Pate agreed.[42]

The other event was beyond WANO's control. With final preparations underway for the Seoul BGM, the terrorist attacks on the World Trade Center in New York City and the Pentagon in Washington, DC, on the morning of 11 September 2001, changed everything. The world watched in horror as television stations broadcast pictures of black smoke and flames rising from one tower of the World Trade Center and a second plane flying inexorably into the second tower and bursting into flames. The attack raised security concerns worldwide and temporarily shut down air travel. With the Seoul meeting just a week away and so many questions remaining about the attacks and the safety of flying, WANO decided to postpone its meeting to the following March.[43]

Although there was some concern among WANO officials that the terrorist attacks might impair WANO's peer review and exchange programmes in the US due to the tightening of access restrictions to nuclear plants, that did not occur. Peer reviews dropped only slightly in 2002 and reached new highs in 2003. Technical support missions and workshops continued to rise as well. Moreover, WANO members rallied behind Atlanta Centre, sending many messages of condolence and support in the wake of the attacks. William Cavanaugh, who represented Atlanta Centre on the WANO Governing Board, said such "expressions of solidarity revealed the strong relationships that have been established worldwide, and were indicative of WANO's success".[44]

The postponement of the sixth BGM provided a few administrative challenges but did not diminish attendance. It also gave WANO time to get its top leadership positions filled. The initial requirement was to reconfirm or replace speakers and panels, and to be certain that speakers updated their talks, taking into account changes that occurred between September and March. The additional expenses resulting from the postponement were approximately $125,000, the costs to be shared equitably among the regional centres. Nonetheless, security at the meeting and on some of the delegate tours was increased due to the September attacks. In addition, WANO asked its members to take additional security precautions to safeguard their nuclear power plants. The impact of the new security measures had little impact on WANO's programmes, however.[45]

More than 345 executive officers and senior executives from every nuclear utility in the world attended the sixth BGM in Seoul in March 2002. In addition, suppliers and manufacturers were also invited, and 17 attended. The delegates unanimously elected Pierre Carlier, the Managing Director for Industry at EDF, as President of WANO. In his opening address to the members, Pate was optimistic about WANO's development and future. He noted the steadily improving performance of the world's nuclear power plants as measured by WANO's performance indicators over the past decade. He also reported that participation in WANO-sponsored workshops and seminars had tripled over the past five years. Pate reminded the delegates of Kupcis's challenge at the Prague BGM in 1997 that peer reviews at all nuclear stations would be completed by 2005. Pate said that "93%...have already scheduled a peer review by 2005. Thus, we are 90% to the goal line with three years plus to go." While Pate was upbeat on the peer review statistics at the BGM, he knew that only plants from the Moscow and Atlanta Centres would meet the 2005 goal. He had learned three weeks before the Seoul meeting that peer reviews would not be completed at all German, French, Swiss and Korean plants by Kupcis's deadline.[46]

This was Pate's last speech as Chairman of WANO. The Governing Board, after interviewing two more candidates, had settled on Hajimu Maeda, an urbane, experienced executive whose career spanned some 40 years at the Kansai Electric Power Company (EPC), the second-largest utility in Japan. Kansai EPC operated 11 nuclear units and provided electricity to major cities such as Osaka, Kyoto, Nara and Kobe. Maeda had joined the company after graduating from the University of Tokyo with a degree in electrical engineering. In the 1970s he had served a five-year term with the World Bank in Washington, DC, financing power projects and providing technical assistance to developing countries. He also honed his English skills during that time, and his facility in the language was a factor in the Governing Board's selection. He returned to Kansai in 1977 to the Office of Nuclear Power Production, rising to Executive Vice President for all nuclear activities. By 1989 he was Executive Director and a member of the Kansai Board. From 1999 to 2001 Maeda served as Chairman of the nuclear power development committee of the Federation of Electric Power Companies (FEPC), in which capacity he promoted the safe operation of nuclear power plants in Japan. He was instrumental in the formation of the Nuclear Safety Network, an industry-wide association of nuclear utilities aimed at enhancing safety culture and improving the performance of its members, much like WANO in promoting peer reviews, information exchanges, and training programmes. His long utility experience, international insight, and active commitment to nuclear safety made him an ideal successor to Pate.[47]

****

Pate could look back over his five-year tenure as Chairman of WANO with satisfaction. Attendance at BGMs had grown steadily, as had the substance and content of the conferences. During WANO's formative years, much of the Governing Board meeting time was occupied in debating various WANO policies. During his first term as

Chairman, the Governing Board hammered out an accepted set of policy documents that "enabled the Board discussions to move forward". Staffing in regional centres had increased, and facilities and programmes in Paris and Moscow had been significantly upgraded. In addition, Pate was particularly pleased with strides made by Moscow Centre to develop peer review team leaders and expand its peer review programme. In 1996, Pate reported, Moscow Centre "was essentially entirely dependent on the Atlanta Centre for peer reviews". By 2001 it was "fully self-sufficient, and their peer reviews are of good quality". The number of technical support missions had grown and the WANO Performance Indicator programme was "solid in every region, used extensively for benchmarking and a real catalyst for improved performance." There was much to be admired in WANO's first decade. No other international industry had voluntarily joined around the concept of self-regulation to achieve operating improvements as had the commercial nuclear utilities.[48]

Nonetheless, WANO was far from perfect. For all of its successes, Pate still heard the echoes of tasks not completed. The rebirth of Moscow Centre was, perhaps, as much a product of Russia's energy policy as WANO's urgings. The Russian strategy to meet domestic energy needs from nuclear power and sell oil and gas to the West, combined with its policy to sell nuclear plants abroad, would fall flat if the country had another nuclear accident. Such economic self-interests made being an active, first-class WANO nuclear safety centre an important part of that energy strategy. Pate had repeatedly stressed this point to the Russian ministers of energy and other top officials. Within WANO, the selection of Pate's successor proved both lengthy and, at times, awkward. Moreover, the Governing Board had grown increasingly concerned, even frustrated at times, when regional self-interests trumped WANO policy and centre directors failed to implement WANO Governing Board actions. The autonomy of the regional centres, once so crucial to the acceptance of WANO in its formative years, was now viewed by many on the WANO Governing Board as outdated and a hindrance to WANO's progress. The casual commitment level of utility executives to the obligations of WANO membership gnawed at the bones of those fully committed. WANO's policies might no longer be subject to monthly Governing Board debates, but the same problems continued to haunt each meeting.

In appreciation for and recognition of Pate's central role in shaping WANO through its first decade and for his leadership as its Chairman, the WANO Governing Board unanimously elected Pate Chairman Emeritus. The action was an extraordinary acknowledgement of Pate's vision for WANO to build a more efficient, effective and robust international organisation, and his strength of character and devotion to achieving that vision.[49] The Board also unanimously approved the establishment of the WANO Nuclear Excellence Award, and a Charter setting out how the WANO programmes would work (see Appendix II).

As Pate left office, the Governing Board additionally adopted a plan that it hoped might resolve some of the outstanding issues. It looked simple on paper. Added to the position of director of the Coordinating Centre would be a second and more powerful position – that of managing director of WANO. However, it would not be so simple in practice.[50]

Chapter Five

THE TRIALS **OF CHANGE**

The search for a successor to Zack Pate had not been smooth. The first three WANO chairmen had been from two regions, Paris and Atlanta. Although the WANO Charter did not require the chairmanship to rotate among the four regions, there was considerable pressure from some members of the Governing Board to select an individual from Moscow or Tokyo. When the leading choice from Moscow Centre imploded his own candidacy, WANO intensified the search. The governors of Tokyo Centre, led by Chair Vijay K Chaturvedi of India, heartily embraced the idea of a chairman from their region. During the summer of 2001, Masateru Mori, a former Tokyo Centre Chairman, approached Hajimu Maeda, who had just retired as Executive Vice President of Kansai Electric Power Company, to ask if he would be interested in the chairmanship of WANO. Maeda was highly regarded in Japan, having occupied several important positions in Japan's nuclear industry, including Chairman of the nuclear development committee of the Federation of Electric Power Companies. As such, he was one of the leaders in the Japanese nuclear power industry, but had little direct relationship with WANO during his years at the utility. He recalled telling his former colleague that he was "not fully qualified for the WANO chairmanship". But Mori urged him to reconsider, and Maeda eventually accepted the offer. With few corporate duties, he decided to take the job.[1]

At the end of October, a month and a half after the terrorist attacks on the World

Trade Center in New York, Maeda flew to London for an interview. The Nominations Committee, William Cavanaugh III and Aleš John, queried him about his views on nuclear power, WANO and the WANO chairmanship. Maeda stressed the importance of nuclear safety, particularly in the volatile post-9/11 world, emphasising that for WANO "there is no goal for nuclear safety where we can stop and we must continue to further enhance our operating performances." Maeda passed the test. In December he received an invitation from Pate to visit Atlanta. For two days the men discussed various aspects of WANO, nuclear safety and Maeda's experience. Of particular interest to Pate was Maeda's scepticism about TEPCO's commitment to nuclear safety in view of that company's falsification of plant data – and the company's major role in determining the policy of Tokyo Centre. Pate believed that Maeda might succeed in making Tokyo Centre more of a WANO team player. For Maeda, the Atlanta visit turned out to be an intensive course in WANO's history, current operations and future direction. Pate also introduced him to Sigval M Berg who, Pate explained, "would be the first managing director of WANO" in London. Fortunately, the two men held similar ideas on the future of WANO. In February the WANO governors unanimously approved a resolution to elevate the role of director of the Coordinating Centre to that of managing director of WANO under supervision of the chairman and the WANO Governing Board. At its March meeting in Seoul, the Governing Board unanimously elected Maeda as Chairman for a two-year term starting in July 2002. The Board also selected Pierre Carlier, a retired EDF executive and former Paris Centre governor, as President and named Berg Director of the Coordinating Centre and WANO Managing Director beginning in October. Personnel to run the organisation after Pate's term ended finally were in place.[2]

Pate had encouraged his close colleague Carlier to become the WANO President. A graduate of the Lille Engineering Institute and Saclay School of Reactor Technology and Engineering, Carlier joined EDF in 1963 and rose in the French utility company's management ranks to become Executive Vice President of Industry before retiring in 2000. Carlier had been a consistent advocate of peer reviews and the WANO goal of conducting one at every plant by 2005. That goal was most important to Pate, who saw in Carlier an ally who – as WANO President – would push his former employer to quit dragging its feet on peer reviews, as well as a champion of completing peer reviews for all of WANO. In leaving WANO, Pate had put in place like-minded successors who he believed could advance WANO's campaign for nuclear safety.[3]

The idea of a managing director had been circulating among WANO's Coordinating Centre staff for some time. Pate, frustrated with the inability of the chairman's office to do more to corral wayward members, looked to provide the managing director power beyond his traditional administrative duties by adding the authority to hold regional directors accountable. Vince Madden, Director of the Coordinating Centre, favoured such a change. Madden's dissatisfaction stemmed from the failure of some regional boards to carry out the directives of the WANO Governing Board if the region disagreed with or did not have the resources to implement WANO policy. The policy divisions between the WANO Governing Board and the regional governors were of long standing. The WANO Charter clearly stated that regional boards could implement WANO policies as they saw fit, which, at times, meant not all. To add to the dissatisfaction in London, the Coordinating Centre's director lacked any authority over the regional directors who decided, in Pate's view, "not to be team players".[4]

Before Madden left in 1999, he and Pate had developed the outline of a plan to strengthen the Coordinating Centre's role. The central concept was to have a more powerful managing director who would get all the regions to act in unison to achieve WANO's goals, particularly regarding the completion of peer reviews. In addition, the managing director would have the authority to write performance appraisals on the directors of the other centres. Since the regional boards would receive these "fitness reports", Pate believed that the appraisal might influence the directors' behaviour to become better team players. In 2001 Pate presented the idea of a managing director to the WANO Governors, where he enjoyed strong support from Governors William Cavanaugh III; Aleš John, who, since 1995, had chaired the Moscow Centre Governing Board; and Stanislav "Stane" Rožman, President of the management board of the Krško nuclear power plant in Slovenia. The latter had been active in WANO since attending the Inaugural Meeting in Moscow in 1989. Rožman's backing was crucial, as he had one foot in the US and the other in central Europe. Because Krško was a Westinghouse pressurised water reactor, Rožman had been active in INPO's International Participant Advisory Committee and, with the creation of WANO, a member of Atlanta Centre. Later, he joined Paris Centre and would become Chairman of its governing board by 2006. Maeda's selection was also part of the equation and sped up the Governing Board's decision. "The idea was to have a seasoned veteran to support Maeda," Pate later explained. For the first time in WANO history, the chairman did not have long-term experience in WANO, and that "was one of the main reasons for establishing the managing director position", Cavanaugh reminded his colleagues. The Governing Board approved the concept and Pate's candidate for the job, Sig Berg.[5]

Berg was Pate's trusted protégé, and at 56 he was 10 years younger than his mentor. A graduate of the US Naval Academy in 1968, he had served under Pate as Chief Engineer of the USS *Sunfish*, where they forged a strong bond of friendship and mutual respect. One of Berg's daughters was named after Pate's wife. Berg left the navy to attend a Lutheran seminary and became a minister to several congregations in the Midwest. He left the ministry in 1988 and went to work for Commonwealth Edison at its Braidwood Nuclear Plant, rising to Site Vice President. In 1994 he joined Pate at INPO, with an eye to working internationally, and was assigned to London to serve as Deputy Director of the Coordinating Centre for a year. During that time

Berg became familiar with the issues facing the director and, from his perspective, the relative impotence of the position. After a year he came back to the US to replace Stan Anderson as Director of WANO's Atlanta Centre. He completed his tour with WANO in 1998 and left for Harvard Business School before returning to INPO. In 2001, at Pate's behest, Berg returned to the UK to head up a WANO review of British Energy – a highly sensitive, complex task, which he carried off successfully. The following year he settled in London as Managing Director of WANO.[6]

Maeda and Berg – the urbane former World Bank officer and senior utility executive and the tenacious former clergyman and nuclear plant operator – made a good team. Maeda had been careful in accepting Berg as Managing Director. Before the Seoul Biennial General Meeting, he had spent considerable time with Berg questioning him about his experience, his views on WANO and the proposed changes in the WANO governing structure. Maeda liked what he heard. Berg thought they "got along very well." Lastly, Maeda turned to Berg. "I have one final question. Who will you be loyal to when you become managing director? To INPO or WANO, the CEO of INPO or me?" Berg replied: "The answer is clear. I report to you. My job is [to] look after and provide leadership for WANO worldwide." The team was in place.[7]

WANO wanted senior utility executives to play a more active role in WANO as well as to reinvigorate the Tokyo Centre members. Maeda exemplified the executive who had participated in other international organisations, including the World Nuclear Association, whose leader, John Ritch, gave him high marks. Although he had had limited affiliation with WANO earlier, he quickly became committed to WANO's programmes and the need to involve more high-profile executives. He clearly saw the importance of reforming Tokyo Centre to bring it more in line with overall WANO policy. Moreover, he was a strong supporter of a more powerful managing director and, by definition, a less direct role for the WANO Chairman. Maeda described his style of directing an organisation as more Asiatic or traditionally Japanese, with the leader setting the policy and strategy by discussion and consensus built among members and a field commander implementing the policy by exercising power delegated to him. Maeda believed that the team of Maeda and Berg worked well "in line with this shared responsibility."[8]

Berg was a tireless worker, winning over his peers with his work ethic and straight shooting more than with personal charm. He was steeped in the safety teachings and techniques of INPO and was part professor and part evangelist, a lecturer and a preacher, when it came to spreading an INPO version of the WANO gospel. Maeda saw himself as a non-executive chairman and valued Berg's capacity for management. Berg relished his new role as Managing Director. For Berg, the position was an opportunity to correct what he perceived as flaws in WANO's governing structure by holding the regional directors more accountable to WANO's policies and by ensuring that all centres followed a standard, coherent policy. "He felt he needed to give better direction and run the WANO organisation in a more consistent way across all four centres," a former associate said. Berg was very focused on what he saw as the path forward. He was a better lecturer than listener, one who would command rather than compromise. Although Maeda was in Tokyo and Berg in London, eight time zones apart, they spoke at least twice a week on the phone, Berg calling in the morning to reach Maeda in the evening. When possible, they blocked out time together, "rich days of conversation, discussion, and strategy," Berg recalled. "It was a solid working relationship with good input." Importantly, both believed a new push for a more centralised WANO with more authority from the Coordinating Centre would strengthen WANO in the future.[9]

Pate was keenly aware that, historically, regional directors and their governing boards had focused their attention almost exclusively within their region. While there had been varying degrees of co-operation and co-ordination between centres, many WANO officials in London and Atlanta were increasingly impatient with a situation that, in spite of the language of the WANO Charter, permitted four regional WANO centres pursuing different agendas with varying levels of commitment instead of one WANO. Therefore, the development of an appropriate balance between a common "global strategy" and "regional/local implementation" had been elusive. The lack of balance between being both an international and a regional organisation, they believed, had impeded the "development and effectiveness of WANO". Maeda, who had witnessed this situation in Tokyo Centre, soon agreed. "WANO must have one global strategy that is culturally sensitive and specifically implemented at each plant/site worldwide [emphasis in original]," Maeda reasoned in his proposal *Moving Forward Together*, which relied on leadership from the managing director and support from the regions in order to succeed. This, he recognised, could only be accomplished through strong and adequately staffed regional centres, strong main and regional governing boards made up of engaged governors who were senior utility executives, and an effective Executive Leadership Team (ELT) consisting of the four regional directors and the managing director. To move forward together, the WANO centres had to work as a team. That meant, Maeda explained, active participation at all levels in WANO programmes, leadership from and cooperation among the regional boards, and the full implementation of WANO strategies by the regional centre directors. The proposal received enthusiastic support from the Atlanta Centre board. The other regions were more moderate in their reaction.[10]

To begin the transition to a new WANO structure and to achieve "One WANO Expressed Regionally" in the late autumn of 2002 Maeda and Berg began a series of visits to each of the four Regional Centres. For Maeda, the visits were *de rigueur*, a traditional tour by each new chairman to introduce himself to each director and each member of the regional boards, discuss WANO programmes and learn the distinct needs and concerns of each centre. Berg accompanied the new Chairman, also doing the meet-and-greet routine. But Berg had the additional task of convincing each of the regional directors that their relationship with the Coordinating Centre was changing. He explained that the centre directors now had two roles – the first as the centre director reporting to a regional governing board, and the second as part of a WANO leadership team that reported to the managing director.[11]

To some regional directors it appeared that Berg wanted to be the chairman of the four centre directors or, at the very least, the centres would be subordinate to him. In addition, from the start some centre directors and regional governors opposed the structural change and the accompanying loss of independence. Nonetheless, Berg plunged ahead, hoping to overcome the opposition. When personality clashes between some directors and Berg became a factor, Maeda and Berg at times found themselves swimming against a stubborn, deep and icy current.[12]

Maeda travelled first to Moscow, visiting Moscow Centre and speaking with Alexander Rumyantsev, Minister for Atomic Energy, who noted "the positive impact of WANO on the improved operation of Russian nuclear plants". He also met with IAEA officials and, while in Paris, with executives of EDF, who agreed that WANO would conduct a corporate peer review of the company in 2003. Maeda also demonstrated that he placed WANO's interests above those of the regional centres. He commented that TEPCO's falsification of records and concealment of plant safety incidents in Japan were "a combined result of inadequate regulations and the operators' slack safety culture". He directly challenged Tokyo Centre's reluctance to conduct peer reviews. The centre's plan for members with more than one station to conduct a peer review only once in 12 years was unacceptable, Maeda said, as it endangered WANO's goal of conducting a peer review every six years. He urged the centre to revise its plans in order to meet WANO's goal, a plea that went unheeded for years.[13]

Shortly after Maeda revealed his tough position for the regional centres to meet WANO guidelines, Berg held his first reorganisational retreat for regional directors, his Executive Leadership Team, in Paris in November 2002. He had selected Paris for two reasons: the first was that the greatest resistance to the position of managing director and oversight from London came from the director of Paris Centre, Yves Canaff; and second was that Maeda and Pierre Carlier, WANO's President and a former EDF executive, would be there to bolster the change in governing structure.[14]

Berg's goal for the ELT was to create a team of five directors to work together to manage WANO. The directors in turn would work with the governors of the regional centres – harmoniously, he hoped – to achieve WANO's goals. Berg would not deal directly with the regional governors; that was to be the role of the centre directors. To illustrate the new arrangement at the meeting, Berg brought in two baseball caps for each centre director. Four caps had WANO emblazoned across the front; the others were each embroidered with PC, MC, AC, or TC to represent the centres. He gave each director a WANO cap and a second one representing their centre, stating that now "you wear two hats".[15]

But baseball was an American pastime and, though embraced in Japan, had limited appeal to other regional centre directors, who had long enjoyed their autonomous position within WANO and saw little value or purpose to tighter direction from London. Only the Directors of the Atlanta and Tokyo Centres, Ed Hux and Yozo Miyazaki respectively, were comfortable wearing both hats. The Atlanta Centre Governing Board favoured any plan for strengthening WANO's operations, but Atlanta stood alone in backing a strong central authority. The Tokyo Centre Governing Board rarely directly opposed a WANO initiative, as saying a direct "no" was considered culturally rude. Rather, regardless of whichever hat the Tokyo Centre director wore, the centre continued to implement WANO programmes according to its own conventions. The Moscow Centre Director, Farit Toukhvetov, "wasn't happy" with the concept, Berg remembered. Canaff was unmoved. He said little, but Berg noticed on his next visit that his WANO hat had disappeared. "I don't like baseball hats, anyway," Canaff explained. Both men had strong personalities, and Canaff fiercely opposed relinquishing any of his centre's independence or, as Carlier put it, he "did not like people crossing his world". Not surprisingly, the relationship between the two men deteriorated thereafter. The attempt to forge a united WANO with the full and unbridled support of the regional centres did not get off to a promising start.[16]

Canaff's chilly regard for Berg and his reluctance to cede the centre's authority to London had a significant impact on the planning of the seventh BGM scheduled to take place in Berlin in October 2003. The theme of the meeting was "Working Together – Safe and Sustainable". It was soon apparent that the theme would not apply to all regions. Traditionally, the Coordinating Centre took responsibility for the technical programme, and the host centre – Paris in this instance, as Carlier was WANO President – was to make all the other local arrangements. According to Berg, Canaff delayed.[17]

As the meeting approached, the number of delegates registered was below expectations. Maeda and Berg urged the regional boards and directors to demonstrate their support for WANO and encourage members to come to Berlin. But Canaff explained that Paris Centre "didn't want to spend a lot of money on the BGM". Berg was appalled how little Canaff had done to prepare for the meeting. "This isn't going to work. Yves doesn't get it," he fumed to Carlier. "I'm not sure he even wants WANO." Berg appealed to Carlier and Laurent Stricker, an EDF executive and WANO Governor from Paris Centre, to replace Canaff. The request, which infuriated Stricker and Paris Centre, "was a tactical blunder on London's part", George Felgate, a later WANO Managing Director recalled, and created "a sore subject for years to come".[18]

Canaff stayed and Paris Centre continued to drag its feet regarding the Berlin BGM. In the past, to take up the slack, INPO would send help. But the new head of INPO, Michael Evans, was less interested in supporting WANO than his predecessors. EDF was also lukewarm to providing resources, and the story circulated that Stricker was supporting Canaff and EDF more than WANO. With no outside help forthcoming, Berg recalled that for the next six months the Coordinating Centre staff, particularly George Hutcherson, worked feverishly on the seventh BGM. In order to promote closer coordination and cooperation within the WANO structure, Maeda invited all the regional governors to attend the WANO Governing Board meeting held just prior to the BGM. The exchange of views among the WANO governors and the regional governors, Maeda recalled, "was lively and fruitful".[19]

The Berlin BGM, which took place against a backdrop of anti-nuclear sentiment in Germany, was a success with 34 countries participating. Openness and transparency were themes of the seventh BGM, and that is what the delegates got – a rousing discussion of existing problems in the nuclear industry. Rolf Gullberg, Chairman of the Paris Centre board and President of KSU, which operated Sweden's nuclear power plants, set the tone of the meeting, attended by nearly 400 members. There had been "a high level of performance in the majority of nuclear plants around the world, but at the same time we have experienced some severe accidents", he told the delegates. Gullberg noted the more serious ones that called into question the industry's commitment to a durable safety culture: failure to identify corrosion on a reactor pressure vessel-head during a peer review at the Davis-Besse plant in the US had badly tarnished INPO's gold standard for peer reviews; and TEPCO's falsification of its plant records had damaged the integrity of nuclear operators everywhere. In April, six months before the BGM, one of the units at Paks in Hungary suffered severe damage to the cladding of some of its fuel rod assemblies while undergoing a cleaning process, with the subsequent release of radioactive gases.[20]

Engineers, it is said, learn more from mistakes than successes, yet those experiences had to be shared and mastered if plant operators were to learn from them; that had not happened. Maeda saw these events as stemming from complacency, the "pit of self-satisfaction", he called it, "a terrible disease [which] threatens nuclear operating organisations from within. It begins with a loss of motivation to learn from others… overconfidence [and] negligence in cultivating a safety culture" due to severe pressures to reduce costs following the deregulation of the electric power market. "These troubles, if ignored," Maeda warned, "can lead to a major accident that will destroy the whole organisation."[21]

Armen Abagyan, a highly respected Russian nuclear official and longtime WANO champion, said that the lack of attention to operational events had contributed to this "new burst of antinuclear opposition and adversely affected the world nuclear industry". The time had come to identify and evaluate the practices that contributed to the effective accomplishment of WANO's mission and pinpoint areas where performance could be improved. To paraphrase Voltaire, "Better is the enemy of good," and good was not good enough.[22]

Berg echoed these concerns, telling delegates that he had identified features common to all these recent incidents. The utilities had accepted degraded conditions – long-standing equipment problems, which individually may not have been burdensome but in combination posed a significant increase in nuclear safety risk. There had been perceived production pressures, "producing a decision-making environment in which risks were not completely understood or considered". Senior management was distracted, often more focused on compliance with licence conditions than safe operations. "There was a failure to be self-critical," which reinforced the message that "problems did not exist or were being adequately addressed". Finally, "most deadly of all", the pretence that plant performance was good "even when this was no longer true". This sense of "goodness", Berg concluded, bordered on arrogance, "creating over-confidence, complacency and in some cases a self-induced isolation from the rest of the industry". WANO programmes, WANO officials emphasised, could be of great value to members, and they encouraged greater involvement. "We are an interconnected international community that depends on each other," Berg reminded delegates. "We cannot and we must not tolerate complacency, arrogance or isolation anywhere in this industry." If members were not actively learning from each other or fully participating in WANO, all should be greatly concerned. That vision set the WANO agenda to strengthen its governance, increase senior executive involvement and improve performance in safety and reliability.[23]

Yet at the time when WANO hoped to increase the commitment of its regional governing board members, the WANO Governing Board failed to achieve a quorum at its first two meetings in 2003. One agenda item, "Moving Forward Together," which was designed to reduce complacency and cost-cutting pressures on operational safety, seemed bogged down even as it started. Maeda believed that there was such a heavy turnover in governors and directors that WANO suffered from a loss of institutional memory or continuity. After Chernobyl there was "great passion" for the concept of WANO, when "world leaders and CEOs demonstrated that support with their personal involvement", Felgate recalled, but by 2000 that "passion had faded, [the relevance] of Chernobyl had faded". Subsequently, CEOs left governing boards and delegated the role of governor down in the organisation, often to a site vice president level. The situation had deteriorated so badly in Tokyo Centre that if the regional governing board made a decision, most governing board members lacked the authority to implement it. To counter this trend, Maeda urged incoming governors to reaffirm the direction of this "new WANO", including members' commitment to the Board of Governors as well as regional directors' commitment to their dual responsibilities – a regional role to manage their centre and a global role to "manage assigned programmes and/or strategic issues". Maeda's comments also marked the beginning of a new concern for WANO – that a younger generation of nuclear operators might not clearly understand or remember the lessons of Three Mile Island and Chernobyl and, as a result, would be less inclined to support WANO.[24]

Maeda was concerned enough with these issues that he queried the Governing Board on its commitment to WANO's new strategy. In summarising the discussion, he noted full agreement that "peer reviews are a most powerful and important tool and need to be improved." He challenged Canaff to lead those improvement efforts with assistance from the other directors. In addition, Carlier emphasised the value of peer reviews – WANO was nearing the completion of its 200th review – urging that the association highlight the benefits of these reviews to the press attending the BGM. Finally, Maeda stressed the Governing Board's agreement that the new unified WANO structure, with the main Governing Board, regional governing boards and the ELT operating in tandem, "is sensible and that strong leadership from the chairman and the managing director is needed".[25]

Maeda believed that the Japanese way of leadership would work "at an international organisation such as WANO which exploits capability at every level" and relies on "a team-oriented way of conducting business". The chairman would rely on his teams and "endorse" their work. Early in his chairmanship Maeda travelled to regional governing board meetings, exchanging views with top executives of major utilities and visiting nuclear plants. "Yet eight to ten time zones between Japan and Europe and America were long legs for travelling," he recalled later, and he was content to let Berg and Carlier do most of the travelling. After an initial series of visits to the regional centres in his first two years, Maeda began to cut back his travel schedule. He would attend the three Governing Board meetings each year but made fewer trips outside Asia. Nevertheless, through frequent telephone conversations, Berg always kept him fully informed of all the activities of the Coordinating Centre. Late in 2003 Maeda was asked to become a commissioner of the Atomic Energy Commission of Japan, the statutory body responsible for formulating the country's long-term nuclear policies. Initially he hesitated about leaving WANO, but "considering the commissioner's important role for Japan's nuclear industry" he concluded that "[I] did not have any option but to accept the position." In December he wrote the Governing Board that he was accepting the Japanese AEC position and would not seek a second term as WANO Chairman.[26]

****

In July 2002 the WANO Governing Board requested self-assessments of each of the WANO centres. A team consisting of the deputy directors from each centre would conduct these internal reviews using a common set of criteria developed with input from each centre. Between March 2003 and February 2004 the team, led by George Hutcherson, Deputy Director of the Coordinating Centre, visited each regional centre, reviewing plans, procedures, reports and other documents relating to the centres, conducted written surveys of centre members and governors, and interviewed centre staff and employees of each centre's members. The final self-assessment reports were to outline the strengths of each centre as well as areas for improvement.[27]

The self-assessments were not harshly critical, though the teams found numerous areas for improvement in each centre. Atlanta Centre, the first to undergo the process, emphasised the importance and value of frequent peer reviews, something non-US members of the centre believed should occur more often than the WANO goal of once every six years. The team found two areas for improvement, both relatively minor. One was to enhance the effectiveness of training peers for the Peer Review programme, an area in which Atlanta had long led WANO, and the second was to be more proactive in obtaining visas for international peers. What team members discovered was more on the minds of Atlanta Centre members, however, was their dissatisfaction with the limited amount of operating experience and other information provided by "some worldwide WANO members". The centre's members were particularly disgruntled because "information on events other than for very high-level, well-publicised events is sometimes not submitted, and in some cases, information on important events meeting the WANO EAR [Event Analysis Reports] criteria is not provided or provided only after significant delay." The report confirmed that most of the "intellectual capital [for events reporting] for WANO came through Atlanta Centre".[28]

The language of the Tokyo Centre assessment was carefully crafted and was "probably not as critical as we could have made it," Hutcherson later recalled. Although the assessment highlighted several strengths of Tokyo Centre, including the use of email and the translation of key WANO documents into Japanese, Chinese and Korean, the primary focus was on the centre's continuing limitations with its peer reviews. "Historically, there has been little follow-up on peer review results to determine if the issues are understood, are being addressed or are of value," something that was "routine" in the other regions, the report stated. Moreover, "important member performance information is sometimes not appropriately discussed or distributed at the appropriate level to allow full use of the information." Tokyo Centre did not share peer review results among the governors, even failing to include the item on the regional governing board agenda and thereby diminishing the centre's ability "to conduct its main job as effectively as possible". The centre's failure to provide sufficient information to its governors and to the centre staff and the lack of interaction with the operating stations, the report stated, "inhibits improving the value of these activities…and inhibits plant performance improvement". A general lack of English proficiency hindered Tokyo Centre's ability to provide peer reviewers

1st Meeting Expert Group IV, 26 January 1988

WANO inaugural meeting in Moscow, 10 April 1989

Nikolai Lukonin and Anatoly Kirichenko at the inaugural meeting in Moscow, 15 May 1989

Anatoly Kirichenko signs WANO charter at inaugural meeting in Moscow, 15 May 1989

WANO's inaugural meeting reception invitation, 16 May 1989

Ian McRae visits WANO Moscow Centre, 1993

Shoh Nasu being elected as the next WANO president to succeed
Bill Lee, 23 April 1991

WANO's 20th Anniversary in St. Petersburg, 20-22 May 2009

George Felgate at WANO BGM 2011, Shenzhen

WANO Internal Assessment Team, 26 November 2012

WANO Moscow Centre at WANO BGM 2013 in Moscow, 19-21 May 2013

WANO **Chairmen**

Lord Marshall of Goring (1989 – 1993)

Mr Rémy Carle (1993 – 1997)

Dr Zack Pate (1997 – 2002)

Mr Hajimu Maeda (2002 – 2004)

Mr William Cavanaugh III (2004 – 2008)

Mr Laurent Stricker (2008 – 2013)

Mr Jacques Régaldo (2013 – present)

WANO **Presidents**

Mr Nikolai Lukonin (1987 – 1989)

Mr William S. Lee (1989 – 1991)

Mr Shoh Nasu (1991 – 1993)

Mr Ian McRae (1993 – 1995)

Mr Erik Pozdyshev (1995 – 1997)

Dr Allan Kupics (1997 – 1999)

Mr Soo-byong Choi (1999 – 2002)

Mr Pierre Carlier (2002 – 2003)

Mr Oleg Saraev (2003 – 2005)

Mr Oliver Kingsley (2005-2007)

Dr Shreyans Kumar Jain (2007 – 2010)

Mr Qian Zhimin (Feb – July 2010)

WANO **Presidents**

Mr He Yu (2010 – 2011)   Prof. Vladimir Asmolov (2011 – 2013)   Mr Duncan Hawthorne (2013- 2015)   Dr Seok Cho (2015-present)

Lord Marshall, Nikolai Lukonin and William S. Lee at the inaugural meeting, 10 April 1989

Robert C. Franklin, First WANO-AC Chairman of Governing Board, 23 March 1989

Zack Pate visiting a pilot peer review in Balakovo, 11-22 October 1993

Laurent Stricker and Qian Zhimin at WANO BGM 2010, New Delhi

Laurent Stricker at Nuclear Excellence Awards 2011 in Shenzhen

Yukiya Amano (IAEA – Director General) and Laurent Stricker sign MOU, 17 September 2012

WANO Young Generation at WANO BGM 2013, Moscow

Jacques Régaldo and Duncan Hawthorne at WANO BGM 2013, Moscow

Panel session discussion at WANO BGM 2013, Moscow

Duncan Hawthorne addressing the WANO BGM 2010 in New Delhi

WANO CEO Ken Ellis visits WANO Tokyo Centre, April 2014

WANO visit to Fukushima, March 2014

WANO Young Generation at WANO BGM 2015, Toronto

Number of events reported to the WANO OE Database

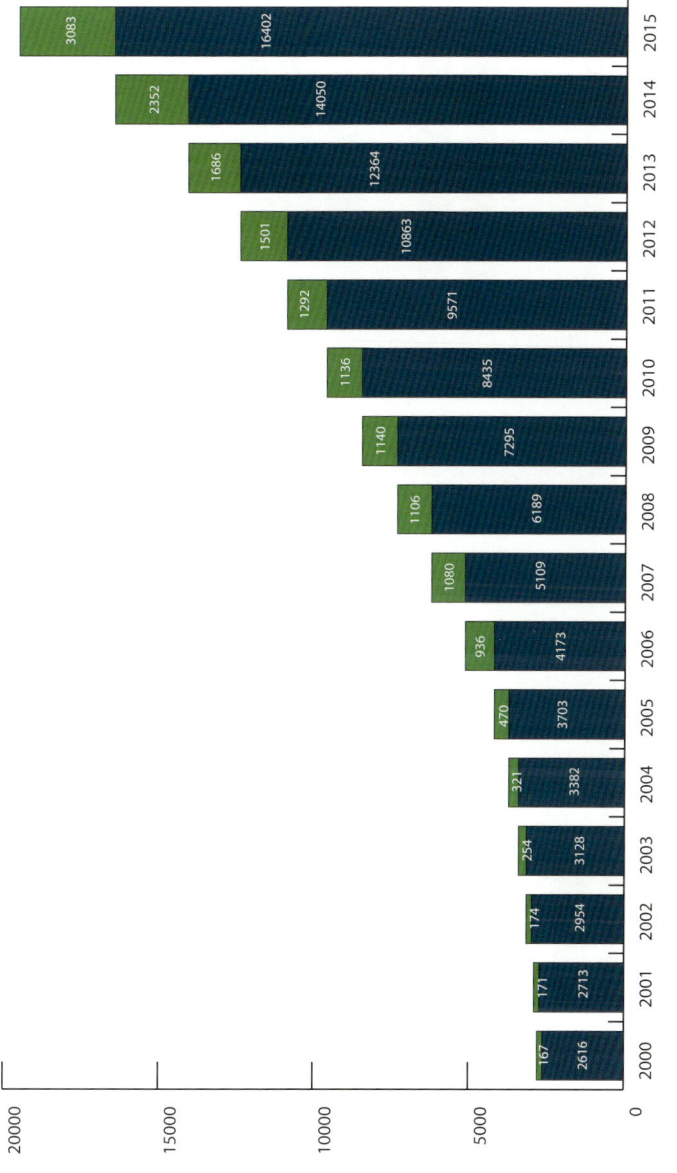

Year	Annual	Previous year cumulative
2000	167	2616
2001	171	2713
2002	174	2954
2003	254	3128
2004	321	3382
2005	470	3703
2006	936	4173
2007	1080	5109
2008	1106	6189
2009	1140	7295
2010	1136	8435
2011	1292	9571
2012	1501	10863
2013	1686	12364
2014	2352	14050
2015	3083	16402

■ Annual
■ Previous year cumulative

The WANO OE Database:
The number of events reported to the WANO OE Database has increased each year since 2000. This reflects an increasingly better reporting culture at our member stations. The cumulative total of reported events from all previous years is shown in navy blue, while the number of events reported in the current year is shown in blue.

WANO PI Trifold 2015 FLR

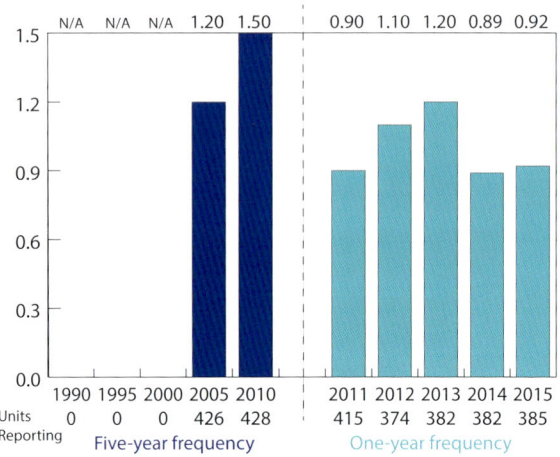

	N/A	N/A	N/A	1.20	1.50		0.90	1.10	1.20	0.89	0.92
	1990	1995	2000	2005	2010		2011	2012	2013	2014	2015
Units Reporting	0	0	0	426	428		415	374	382	382	385
			Five-year frequency					One-year frequency			

Forced Loss Rate (FLR)
The forced loss rate is the percentage of energy generation during non-outage periods that a plant is not capable of supplying to the electrical grid because of unplanned energy losses, such as unplanned shutdown or load reductions. A low value indicates important plant equipment is well maintained and reliably operated.

WANO PI Trifold 2015 ISA

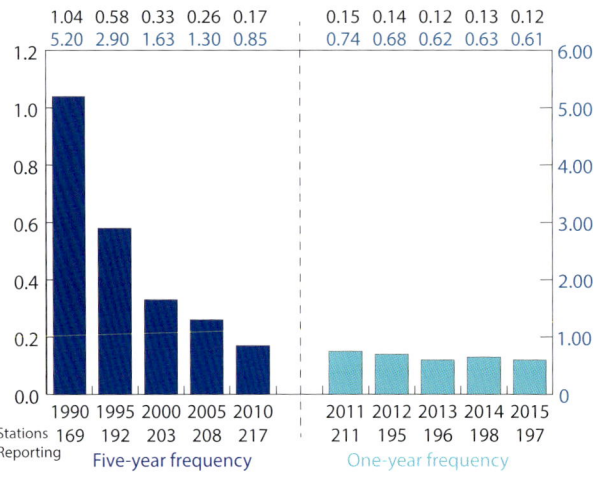

	1.04	0.58	0.33	0.26	0.17		0.15	0.14	0.12	0.13	0.12
	5.20	2.90	1.63	1.30	0.85		0.74	0.68	0.62	0.63	0.61
	1990	1995	2000	2005	2010		2011	2012	2013	2014	2015
Stations Reporting	169	192	203	208	217		211	195	196	198	197
			Five-year frequency					One-year frequency			

Industrial Safety Accident (ISA)
The industrial safety accident rate tracks the number of accidents among employees that result in lost work time, restricted work, or fatalities per 200,000 work-hours (and 1,000,000 hours worked). The nuclear industry continues to provide one of the safest industrial work environments.

WANO PI Trifold 2015 US7 and UA7

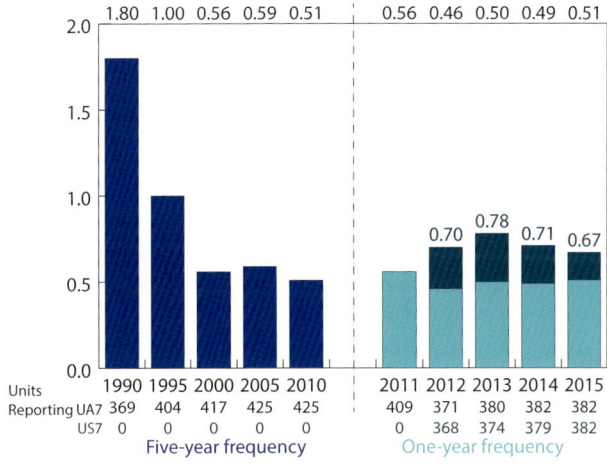

Units Reporting	1990	1995	2000	2005	2010	2011	2012	2013	2014	2015
UA7	369	404	417	425	425	409	371	380	382	382
US7	0	0	0	0	0	0	368	374	379	382

Five-year frequency | One-year frequency

The unplanned automatic scrams per 7,000 hours critical indicator tracks the mean scram (automatic shutdown) rate for approximately one year (7,000 hours) of operation. Unplanned automatic scrams result in thermal and hydraulic transients that affect plant systems. US7 data collection began in 2012 and from this point UA7 (light blue) is displayed as a part of US7 (unplanned total scrams).

WANO PI Trifold 2015 UCF

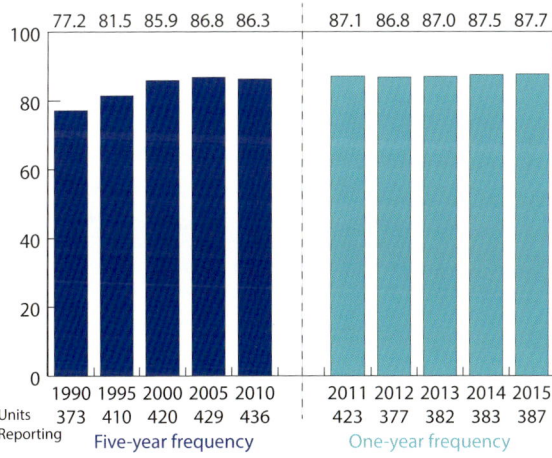

Units Reporting	1990	1995	2000	2005	2010	2011	2012	2013	2014	2015
	373	410	420	429	436	423	377	382	383	387

Five-year frequency | One-year frequency

Unit capability factor is the percentage of maximum energy generation that a plant is capable of supplying to the electrical grid, limited only by factors within the control of plant management. A high unit capability factor indicates effective plant programmes and practices to minimise unplanned energy losses and to optimise planned outages.

WANO PI Trifold 2015 UCLF

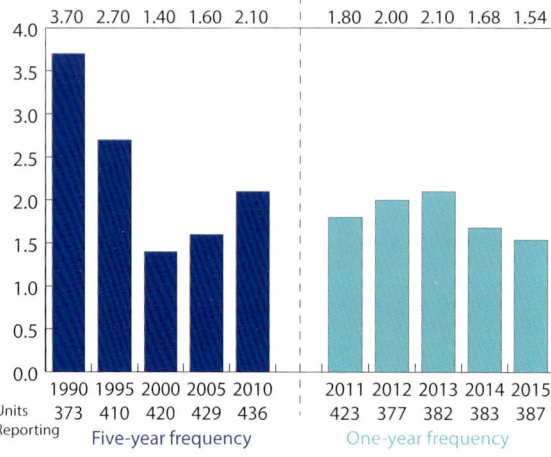

Unit capability factor is the percentage of maximum energy generation that a plant is capable of supplying to the electrical grid, limited only by factors within the control of plant management. A high unit capability factor indicates effective plant programmes and practices to minimise unplanned energy losses and to optimise planned outages.

with the necessary expertise and experience, and regional training programmes were not considered sufficient by members outside of Japan. Without stating it directly, the report indicated that Tokyo Centre was more an adjunct to WANO than a full partner. While the report said that peer reviews had revealed "a high number of strengths" and that many "Tokyo Centre stations have very good performance results" the centre itself identified or shared few of these "many potential good practices", as was the practice in other regional centres.[29]

Its members cobbled together from all of Asia, Tokyo Centre had since the beginning of WANO been the most difficult region to manage. Its governing board was composed of few senior executives. Japanese CEOs had usually delegated assignments to the regional governing board to lower-level officials, who had to take back any board decisions to their respective organisations for approval, a time-consuming and uncertain process. The language barriers among its members were more severe than those faced by any other centre. Cultural differences and the wounds of World War II contributed to a lingering distrust among Tokyo Centre's members, as did the uneasy relationship between India and Pakistan. Such differences made reaching a consensus particularly difficult. WANO and Tokyo Centre had considerable hurdles to overcome on the road to improvement.[30]

The assessment team moved on to Moscow Centre. WANO had invested a great deal of money and resources into the Moscow operation and, by the first years of the 21st century, was beginning to see some returns. The report noted the "proactive efforts by the centre to involve all members in WANO activities" and its "continued aggressive action to conduct and improve peer reviews". Since WANO's founding, Moscow had been preoccupied with resurrecting, repairing or retrofitting its Soviet-era plants. After so many years of being propped up by other centres – with money, manpower or both – Moscow Centre was improving. The assessment team was particularly impressed with the centre's outreach to each member, scheduling some WANO activity at each plant during the course of a year. Moreover, the Centre had improved its secure website, despite some difficulties with encryption issues in some member nations. Rather than focus on keeping plants in operation or implementing WANO programmes, the report recommended that the centre be more focused on improving performance issues identified in peer reviews in areas such as radiological protection and fuel performance. As it had with Tokyo Centre, the assessment team urged Moscow to conduct more follow-up reviews to ensure that plants had taken the corrective actions recommended in peer reviews.[31]

The tenor of the recommendations for improvement reflected the enormous distance that Moscow Centre had travelled. Wary at first of the West's strictures on achieving safe operations, the members of the Moscow Centre now embraced its activities. Importantly, the assessment team reported a high level of support among all the Centre members for WANO's Peer Review programme. The members also expressed finding great value in WANO's technical support missions, even though most of those missions went to other Moscow Centre plants and there was little sharing of resources across other WANO regions. Moscow Centre, so crucial to the success of WANO, had finally begun to demonstrate WANO's value to operational safety – from the centre most in need to a centre committed to WANO's mission.[32]

What the assessment report did not mention was the reluctance of the Moscow Centre to deal forthrightly with the 2003 accident in which fuel bundles were damaged during cleaning at the Paks Nuclear Power Plant. Neither the Russians nor the Hungarians would permit a WANO team to investigate. WANO persisted. When a team finally gained clearance, it was welcomed by the station's employees, who were quite open about the event. But there was "no ownership" or support from the top, one team member recalled. "There was a really poor safety culture". Paks had been the first European plant to undergo a WANO peer review in 1992. Its operating record had been good; no further peer review had been done because "there were so many other things WANO needed to do and so few resources" with which to do them. The critical issue, according to one WANO veteran, was that the event demonstrated an abdication of ownership for the fuel and nuclear safety once the fuel was turned over to a contractor for cleaning. At the eighth BGM in Budapest in 2005, the Hungarians apologised to the members for fumbling on nuclear safety. The lessons for WANO were (1) regional centres would be protective of their plants in the event of an accident rather than following WANO procedures; (2) WANO should not assume plants continued to operate well after a peer review; and (3) a set schedule of more frequent peer reviews needed to be established.[33]

The review of Paris Centre also concentrated on peer reviews. The assessment team praised the centre for its preparation for peer reviews, the formal analysis of performance indicators, operating experience data collection and analysis, and the conduct of pre-visit interviews. But thereafter the process broke down. Paris Centre did not meet WANO's peer review goals: its peer review reports lacked timeliness and were of "inconsistent quality" and the centre had difficulty obtaining experienced peers to conduct the reviews. Some of the centre's members, the report stated, "demonstrated a lack of full commitment to WANO". In addition to the centre's casual approach to completing peer reviews, the team zeroed in on the fact that "important policies, decisions, information and expectations communicated by centre management are sometimes not fully understood by the centre staff." This was the team's polite way of saying that the Paris Centre was not fully supportive of WANO's programmes.[34]

But that lack of support was echoed throughout the region. Part of WANO's frustration with the Centre was that some of its members did not make WANO a high priority. One Paris Centre governor believed that senior utility executives, most of who did not participate directly on the Paris Centre Governing Board, "think they do not need WANO, as performance in Europe appears to be very good". EDF and the German utilities certainly shared that belief, one that was also part of Tokyo Centre's culture, particularly represented by TEPCO. It was for this reason that Maeda and Carlier, who had become special adviser to the chairman when his Presidency ended, preached against complacency and over-confidence in the industry. As Abagyan once warned, "When you only have success, it is a danger because you lose attention. When there is success after success, be careful."[35]

The last centre to undergo self-assessment was London. The review concentrated on the secretariat side of the centre – the conduct of board meetings and support to the chairman, governing board, directors' meeting and communications/publications – rather than the change to a managing director or the move of the Operational Experience Central Team (OECT) from Paris to London, both of which Paris had opposed. The review team found that because of the structural change, "some roles, responsibilities and relationships need to be more clearly articulated and defined." One major area for improvement was staffing. It was a long-standing concern as regional centres, with the exception of Atlanta, had always been reluctant to send their best people to work in London or to readily fill vacancies when positions opened. The problem was one of resources and perception. Atlanta Centre had resources and saw great value in WANO and in its success and survival. It generally sent highly competent staff to London. The other centres "believed London added very little value; all programmes were run independently by the regions and London just added bureaucratic overhead," according to one WANO official.[36]

While the centre self-assessments varied in the strengths and weaknesses of each centre, there were several priorities common throughout WANO. One was that WANO had to improve CEO involvement and regional support for the organisation. WANO also needed to place additional emphasis on direct support to its members to correct performance deficiencies. In addition, the need was readily apparent for increased numbers and more experienced staffing in the five centres, as well as more continuity of staff that might be achieved by lengthening the assignment terms of seconded, or loaned, employees. The current situation "continues to be an impediment to WANO reaching its full potential".[37]

The timing of the self-assessment reviews was ill-fated. The Governing Board, faced with what it deemed more pressing issues, permitted many of the recommendations – deliberately written to avoid controversy – to fade into the background. The core issues did not recede, however, and two years later would become the focus of the next WANO review. The association continued to work toward the self-assessment team's recommendations, and some of the gaps identified in the assessment were partially closed. Yet the gaps consistently gnawed at WANO's leadership, and it would be nearly another decade until greater numbers of senior utility executives sat on regional boards. As the centres expanded, vacancies in the Coordinating Centre were filled, and the quality of the staffing throughout all the centres improved.[38]

The Governing Board's more immediate and troubling issue was the deteriorating working relationship between the managing director, directors and regional governors. The centre reviews had exacerbated suspicions that the Coordinating Centre might use the assessments to reduce regional autonomy. At a meeting in January 2004, Maeda and the Governing Board put WANO governance at the top of their agenda. The relatively high turnover of governors and directors had led to misunderstandings regarding the governance of WANO, the Governing Board believed, and "key roles and relationships needed to be reaffirmed". The Governors agreed that there was a clear line of responsibility from policy-setting by the WANO Governing Board to implementation by the managing director through the regional centres. In addition, cooperation and coordination were required to implement those policies. Yet that was not the case, Maeda protested. He stressed that WANO had a common strategy and "each region cannot go in its own way". Nevertheless, he added, "recently, there have been some indications of this."[39]

Contributing to the problem was the feeling in most regions that the Governing Board often made policy decisions without getting sufficient input from the regional boards. In the main Governing Board meetings, topics would come up without giving the regional representatives an opportunity to discuss them among their members. The meetings, according to one observer, were often dominated by Atlanta and London, and Tokyo Centre representatives "would sit quiet".[40]

Nevertheless, the issue was real. Carlier confirmed Maeda's view. He had attended a recent directors' meeting, observing that while "the directors spoke about co-operation, [they] clearly did not cooperate". He backed Berg's efforts to lead the ELT. "Regional directors need to be managed," he said, "not just coordinated." The managing director, the Governing Board decided, should "ensure proper organisational alignment of the Executive Leadership Team with the WANO Governing Board." But the Governing Board agreed that it, too, had contributed to the problem. Communication between the WANO Governors and the regional governing boards and directors was "a weak area" and needed improvement. The Governors also revised the governance documents to make clear that the managing director review of regional directors was in respect only to their ELT and cross-regional, or WANO, responsibilities. By giving the managing director a title without the authority to match, the Board, in its indecision, also contributed to Berg's failure.[41]

The second issue was nominating Maeda's successor. At the beginning of 2004, Maeda had taken up his commissioner's position and was spending most of his time in Japan.

The Governing Board wanted to act quickly and elect a candidate by its April meeting in Nara, Japan. Maeda appointed Oliver D Kingsley, Jr, (a new member of the WANO Governing Board and President and CEO of the Exelon Corporation headquartered in Chicago), and Aleš John to be the Nominations Committee. Cavanaugh lobbied for a new chairman who was heavily involved with WANO, preferably someone with governing board experience. He also emphasised that the Nominations Committee should name candidates willing to take the position, and the candidates should be from a company that would provide adequate support to enable them to serve. His suggestions indicated that he wanted WANO to elect a stronger, more active chairman in the tradition of a Marshall, Carle or Pate. According to the minutes of the meeting, Cavanaugh did not state whether he himself would be a candidate for the chairmanship or not.[42]

Maeda asked the Nominations Committee to select only one candidate. When the Governing Board met in Nara in April, the nominee was Cavanaugh. Immediately, the choice generated considerable controversy. The Paris and Moscow Centres, both of which had put forward candidates of their own, were united in their opposition to an American chairman and an American managing director. Moscow Centre pushed for a candidate from its region, outgoing WANO President Oleg Saraev, who was organising the 2005 BGM in Budapest. But because Saraev came to meetings with an interpreter, the Governing Board did not believe he was sufficiently fluent in English to handle the job.[43]

Other potential candidates from the region were plant managers, not the senior executives that the Governing Board thought should lead WANO. In Nara, Paris Centre urged the Governors to consider Laurent Stricker, chair of the centre's regional governing board, a senior executive at EDF and a fluent English speaker. But because of his support of Canaff and lukewarm backing of peer reviews, Stricker was viewed as an EDF man rather than a WANO stalwart. The governors debated the matter for most of a day. In a split vote, Cavanaugh won. Yet Americans could not hold the top two positions in WANO. With Cavanaugh's selection, Berg offered his resignation. He would serve out his two-year term, to be replaced by Lucas "Luc" Mampaey, a Belgian who – as a Francophone – might repair the strained relations between the Coordinating Centre and Paris Centre.[44]

In a letter to the WANO Board of Governors announcing his departure, Maeda urged WANO to shift from being programme-oriented to performance-oriented. He praised WANO members for their "keen attitude for safe operation". WANO should be "strongly focused on improving the performance of our member plants by providing tailor-made services for each individual plant with a diagnosis about performance for each plant". For WANO to move forward, he said, "it is necessary to reform WANO". He invited the Governing Board to put policies into place to increase staffing, improve communications between the Governing Board and WANO members and revitalise regional centres. "Openness and sharing among members needs to be enhanced," he stated. Years later, in summing up his term as WANO chairman, Maeda emphasised a number of principles he had backed that would make the organisation stronger in the future. Among those measures to achieve WANO's long-term goals he included the expansion of WANO's Operating Experience programme, the promotion and development of peer reviews and the expansion of membership to accommodate vendors, nuclear fuel fabricators and processing facilities. He was proud of the initiative to involve nuclear operators more deeply in WANO activities and provide the encouragement to maintain plants in good operating condition and at high levels of safety. He also had presided over a pivotal period of change in WANO governance. Without Maeda's full support of the shift in leadership to a managing director heading the Coordinating Centre and open discussion among WANO and regional governors, the future of WANO might have been very different. Although the experiment of a "One WANO" gained little traction at the time, it would have far-reaching implications in the future.[45]

****

The change in WANO's leadership signalled a shift from the strategy of a powerful managing director as Berg envisioned it. Cavanaugh saw himself in the Pate tradition of a strong chair – an executive chairman – and quite different from Maeda who was willing to let the managing director take the lead. Under Cavanaugh, Mampaey's role as managing director became less managing and more administrative, similar to that of the Coordinating Centre directors who preceded Berg. But the change in top personnel in London did not signal a shift in the basic strategy to strengthen the authority of the Coordinating Centre and bring the regional centres and their boards into full alignment with WANO's policies. Cavanaugh would take the lead. Accordingly, even with Berg out of the equation, the drive towards One WANO continued.

Cavanaugh had a long pedigree of WANO participation. Like Berg, he was a Pate protégé and avid admirer. Born and raised in New Orleans, Cavanaugh attended Tulane University under a Navy ROTC scholarship, graduated with a degree in engineering in 1961 and then, to fulfill his scholarship obligation, joined the US Navy's nuclear programme. After eight years he left the nuclear navy and joined Middle South Utilities, later Entergy, which operated several nuclear plants. After Chernobyl, he and Pate discussed the formation of a new international organisation along the lines of INPO. Cavanaugh expressed his interest in its formation and attended the planning meeting in Paris in October 1987. Years later he recalled how impressed he was with the utility executives, particularly Bill Lee, Lord Marshall and Rémy Carle, who worked tirelessly at that initial meeting to convince the Russians, the Japanese, and the French to support a new international organisation.[46]

Energetic, even aggressive, leadership from senior executives remained a crucial element in his thinking regarding a successful WANO. While in Europe for the Paris planning meeting in 1987, Cavanaugh visited plants in Switzerland and Germany. He was impressed that they refuelled in half the time it took Americans. He brought the European operators to America to demonstrate how they did things more efficiently. The experience convinced him of the value of exchange visits and international cooperation. He also attended the first meeting in Moscow and was one of the original signatories to the WANO Charter. He progressed quickly up the Entergy ladder, moved to become President and CEO of Carolina Power & Light (CP&L) Company, then became President and CEO of Progress Energy in 1999 until his retirement in 2004. In 1992 he replaced Pate as one of Atlanta Centre's members on the WANO Governing Board. Cavanaugh remained a governor until he replaced Maeda as Chairman in July 2004.[47]

Ten years Cavanaugh's junior, Mampaey was a graduate of Louvain University with a degree in electromechanical engineering and an advanced degree in nuclear engineering. After graduation and army service, in 1973 he joined Tractebel, a Belgian engineering company conducting the design and system engineering for the initial two nuclear units being constructed at Doel for Electrabel. By 1986 Tractebel had merged with Electrabel, and Mampaey was working on the design of a fifth reactor at Doel. "We had everything ready, all the systems descriptions, all the flow sheets. We were ready to order equipment. Then came Chernobyl. Everything we did went into the wastebasket."[48]

With no more reactors likely to be built in the foreseeable future, Mampaey concluded that nuclear construction engineering was not the place to be, so he switched to operations and rose to become Plant Manager, then Site Vice President at Doel. In 2004 he became a regional governor at Paris Centre. He saw himself as a champion of WANO and of nuclear safety, but he was also aware that WANO was changing – and not necessarily for the better. "Chernobyl was already long forgotten"; its urgency had "faded away", he recalled. When he learned that Berg was leaving, he saw an opportunity to go beyond Electrabel and remain in the nuclear industry. For someone who had spent his career in Belgium, the international aspect of the managing director post appeared most attractive. Mampaey offered his candidacy and was selected.[49]

Cavanaugh and Mampaey encountered a number of changes that would affect WANO's operations. The process of deregulation and competition in the electric power industry in Europe and the US had a telling impact on budgets. For utilities that believed they ran plants better than elsewhere in the world, WANO's mission seemed less important, so why should they spend money on it? Under new CEO Michael Evans, INPO support after 2002 became lukewarm at best. Evans believed that WANO was a burden on INPO's resources and did not provide commensurate benefits for INPO members. He was far less willing than his predecessors to approve the use of INPO personnel or travel funds for WANO peer reviews and assistance at international nuclear sites. As a result, Atlanta Centre lost some standing with its international partners.[50]

The Paris and Tokyo Centres encountered similar reactions. According to Mampaey, when he arrived in London, EDF and the German utilities saw more value in their procedures and believed they gave more to WANO than they received. This attitude, he thought, was also reflected in the failure of the WANO Governing Board's policies to be implemented by Paris Centre. In addition, British Energy was undergoing severe financial difficulties, straining WANO's resources. Mampaey also bemoaned the lack of top executives on regional boards. The level of governors sent to the two regional boards, he said, "was mixed", with more plant managers and fewer CEOs holding those positions. Moscow Centre was just recovering from its decade of financial troubles, and the centre's budget problems were still exacerbated by members who would not, or could not, pay their WANO fees. In Moscow, too, governance came from plant managers. Mampaey saw all of these issues reflected in the lack of participation in WANO programmes and the lack of use of WANO's products. There were too many differences between the centres, too many plants without peer reviews, and no follow-ups for many that had undergone a peer review. There were no repercussions if the problems found in a WANO peer review went uncorrected. Mampaey likened WANO to a golf club. "You come to play when it suits you," he said.[51]

Yet in spite of the bad news, other utility executives saw a silver lining: the possibility of a renaissance in the nuclear industry, not only in America but in Asia, particularly China and India. Cavanaugh and Mampaey intended that WANO and its Governing Board would play a major role in any nuclear renaissance and to resolve WANO's internal problems as well as strengthen programmes to improve plant safety and performance. Beginning in the autumn of 2004 and for the next several years, the two men, with help from Carlier, who remained a special adviser to the chairman and an active advocate for greater member involvement and responsibility, called for more frequent peer reviews, urged the upgrading of WANO's programmes, and pushed centre directors and regional governing boards to bend to a central authority. Several Governing Board members fully backed the chairman's efforts, particularly Kingsley and W George Hairston III, President and CEO of the Southern Nuclear Operating Company and newly elected chairman of the board of the Nuclear Energy Institute (NEI), from Atlanta Centre, and Aleš John from Moscow Centre. However, other WANO governors, who represented regions that wished to preserve their autonomy or particular definitions and implementations of WANO's policies, dragged their feet. The struggle between the two factions – those who sought a more centralised and uniform WANO and those who fought to keep the status quo based on the original

WANO Charter and "cultural differences" – would continue to play out over the next decade, with the future direction of WANO at stake.[52]

Improving WANO peer reviews became Mampaey's first challenge. The governors thought WANO should conduct more frequent and more effective peer reviews. The regional self-assessments demonstrated a weakness in implementation of the Peer Review programme for all the centres with the exception of Atlanta, which provided the bulk of WANO's peer review teams. Increased peer review training was an essential aspect of "beefing up" the programme, so the Coordinating Centre rewrote the criteria for evaluator training and development of more team leaders. Mampaey also sought to increase staffing at the centres. Atlanta had normally been fully staffed but had received little support from the INPO CEO after 2002. With Evans's departure in 2004, Fred Tollison, another Pate protégé, took over and restored INPO's previous support of WANO. Atlanta was the exception, however, as the other centres were woefully understaffed. Hutcherson, who returned to London to serve as Mampaey's deputy in 2004, estimated that London had a staff of eight or nine, Paris 20, Tokyo and Moscow about 12 each – in all, about 50 people to run an international organisation. Staffing, too, would require attention.[53]

In the spring of 2005 at a meeting in Barcelona, Spain, the WANO Governing Board, after some discussion, approved a Long-Term Plan to provide overall direction for the subsequent three years. The Plan re-emphasised the dual responsibilities and obligations of WANO members: that plant operators had "**both** the individual responsibility for nuclear safety **and** a mutual or collective responsibility for nuclear safety of operating nuclear power plants world-wide [emphasis in original]". The Plan outlined three major goals – to support each member in improving the safety and reliability of nuclear facilities, for the WANO membership to meet its collective responsibility to improve performance and "continually upgrade the safety of all nuclear plants" and to maintain an organisational support structure, staffing, and membership so that "WANO can consistently work effectively in a changing environment". The most effective way to achieve these goals, the Plan suggested, was to conduct more frequent independent peer reviews by well-qualified and experienced team leaders and members, increase member participation in WANO programmes, and "adjust", or increase, the quality and professional standing of WANO staff.[54]

The nature of peer reviews had been at the heart of the Governing Board's discussion. The Americans and Carlier pushed for a greater number of independent, or outside, peer reviewers, a position not favoured by the Japanese or French, who jealously guarded their own approach to peer reviews. On top of that Tokyo Centre faced considerable language barriers, having few experienced English speakers. In addition, India, which had considerable resources, could not participate in many peer reviews due to its stance on the nuclear non-proliferation treaty. Carlier opined that EDF reviews lacked "a non-EDF perspective" and that outside reviewers would be a valuable addition. Cavanaugh agreed, and the Governing Board acceded to compromise that peer reviews include "some percentage of reviewers [from] outside the utility". The Governors left it to the ELT to work out what that percentage might be. Another sticking point was the conduct of corporate peer reviews, which under the 2005 Long-Term Plan were to be determined by the WANO Board of Governors and coordinated by the managing director. The Paris and Moscow Centres disagreed. The responsibility for corporate peer reviews should "reside with the appropriate regional centre", Canaff and Toukhvetov asserted, and not with the managing director. Canaff further said that he was unclear what "coordination by the managing director" meant. Their opposition killed the centralisation of the corporate peer review process. The Governors decided that the responsibilities for corporate peer reviews would reside with the centre in which the utility was located.[55]

The updated Long-Term Plan was part of a campaign conducted by Cavanaugh and his allies on the Governing Board to strengthen the operational and governing structure of WANO. To achieve this effort, Cavanaugh sought greater utility support in the Paris Centre region. He viewed the turnover of WANO governors as an opportunity to indoctrinate new Governing Board members to help in the transition. He pushed for higher quality of peer reviewers. In addition, regional directors were to identify weak-performing plants so that WANO could provide appropriate assistance and develop approaches to deal with those plants that did not improve over a long period. The identification of poorly performing plants followed by swift WANO assistance would be initial steps in providing WANO with more authority to deal with safety concerns correctly, before an accident occurred. To be certain that regional directors followed the directive, the Governors asked the Managing Director "to report how the ELT is functioning" at subsequent executive sessions of the Governing Board. Mampaey's other assignment as part of the offensive to enhance WANO's governance was to recruit utility executives to become more involved with WANO. The idea was also to urge the utilities to reaffirm their commitment to WANO. In the spring of 2005 he visited key executives in Europe and expanded his trips into other regions during the year.[56]

Shortly before the 2005 Budapest BGM in October, the Russians, led by Oleg Saraev, WANO's President, joined the effort to tighten WANO's governance and work toward a common vision for the association. Saraev explained that, historically, WANO had relied on the goodwill of members to make improvements, but this had not always occurred. Looking back over WANO's evolution, Carlier noted that "we have achieved a lot in 15 years, but we could have achieved more." Saraev stated that "now is the time to consider formal agreements between the WANO member plants and WANO to clearly delineate responsibilities" among the plants, their regional centres, and the Coordinating Centre. He hoped that such an agreement would encourage plants to use more WANO products and to better define "the WANO-utility interface and obligations".[57]

Saraev's appeal touched a lingering concern among the WANO Governors that WANO needed to "solidify its worth to [its] customers…and target plants that do not participate much". Moreover, some Governors saw it as an opportunity for WANO's "key leaders" – members of the WANO and regional governing boards – "to rise above divisive issues". Saraev's proposal quickly won approval from his two Moscow Centre colleagues, Yuriy Nedashkovskiy and Aleš John, and from Hairston and Kingsley. The governors from the Paris and Tokyo Centres said little but did not oppose the recommendation.[58]

Saraev's proposal meshed perfectly with Cavanaugh's agenda for WANO. At the organisation's eighth BGM in Budapest, Cavanaugh warned that the level of involvement and commitment to WANO had "to go to the next level". On the eve of the 20th anniversary of Chernobyl, he told the members that the rapid improvements that characterised WANO's first decade had stalled, or even declined, in some areas. He particularly pointed a finger at utilities with a large fleet of plants that thought their experience base was sufficiently large so that outside contact was not important. Though he did not mention names, the German, French and Japanese utilities were widely regarded as believing they operated better than others and that WANO's programmes held limited value for them. "This is the same mentality of self-sufficiency that existed before WANO's formation," Cavanaugh warned, "and we all know the cost of isolation is high." From a financial perspective, he said, there was no better insurance policy against poor performance "than participating fully in WANO programmes".[59]

Cavanaugh's recommendations suggested adopting a more INPO-like activism. All levels of staff must participate, he said. If WANO were to deliver better products for its members, utilities needed to send talented staff to the centres for longer periods of time. A WANO assignment should be seen as a step in career development for rising managers, not as a short-term post or a sinecure for retiring employees. Moreover, participation at the highest level was vital. "A CEO who is visibly committed and active in WANO sends a clear unambiguous message about the importance attached to nuclear safety." To help ensure that a member was responsive to WANO's obligations, he followed the pattern set by Pate in dealing with Chernobyl and its operator Energoatom, calling for intervention in the event of the plant being unwilling to correct problems or having "persistent shortfalls in performance". When all else failed, under these "special conditions" the WANO chairman would contact the member's CEO and, ultimately, its board of directors to resolve the situation. By outlining what would happen to a recalcitrant plant as part of WANO's obligations, Cavanaugh sought to put the teeth of enforcement into WANO's membership policy.[60]

To eliminate what he called "barriers to sustained WANO effectiveness and continued improvement", immediately after his re-election as WANO chairman in 2006, Cavanaugh appointed a Special Committee to review "WANO processes, programmes, activities and organisational relationships." The Committee was to consider the future role of WANO and make recommendations that would help shape the organisation for the years ahead. Cavanaugh told the Committee to start with "a clean piece of paper" and make recommendations so that the Governing Board "can have a shared vision and a clear understanding of our organisation and leadership succession". He appointed Oliver Kingsley, WANO's new President, to head the committee and asked the chairman of each regional governing board to be a member. Kingsley, a recipient of a WANO's Nuclear Excellence Award in 2003, was an ideal choice. He enjoyed widespread respect for turning around poorly performing plants for the Tennessee Valley Authority and Commonwealth Edison in the US and for his vigorous commitment to nuclear safety and operational excellence. He was a strong leader who demanded results – and got them. Those who did not perform were dismissed. However, applying Kingsley's experience to an international organisation did not ensure the same turnaround success he had enjoyed in North America.[61]

The Committee's charge included many of the issues and concerns that the WANO chairman and Governing Board had confronted for years and, accordingly, would draw on the self-assessment review conducted two years earlier. Everything would be on the table for discussion. The Special Committee would reexamine the WANO mission, which established the voluntary exchange of operating experience as its primary activity. Should the WANO mission be revised so that it "would be a strong forcing function for plants to improve performance – not just a facilitator of information exchange?" he asked. Was an improvement in governance structure needed? Should there be changes in the executive and leadership authority? He asked the committee to consider if the current staffing arrangement (primarily seconded staff with frequent turnover) was suitable for the future in regard to staff quality and continuity.[62]

The Committee began its work as the landscape of the nuclear power industry had started to shift. As part of the nuclear renaissance, new plants were being constructed across the world, new countries were becoming nuclear operators and large nuclear companies such as EDF were forming and becoming global operators. The expansion of nuclear plants had "exerted considerable pressure on human and technical resources", Cavanaugh said. Many new utility CEOs had limited or no experience with nuclear plants. At the same time, performance was tailing off as many nuclear plants began to show their age while they edged toward the end of their designed lives and decommissioning. With all this in mind, the Special Committee began its work.[63]

What the Committee did not know as it began its fact-finding research was that a serious event had occurred on 1 March 2006 at Bulgaria's Kozloduy plant. More than a third of the control rods could not be inserted into the core. Moscow Centre knew of the accident but did not say anything to its WANO colleagues at the Governing Board meeting in Beijing in April. The Bulgarians, looking to join the European Union, refused WANO's request to send a team to visit the plant. Moscow Centre, which held that it was unnecessary to report to WANO if there were no consequences from the event, downplayed the accident and would provide only limited information to WANO. Moscow's reticence, some WANO officials surmised, was partly because a similar Russian-built plant, Tianwan, in China had experienced the same event. Releasing information on Kozloduy might further unnerve the Chinese and damage opportunities for future nuclear plant exports. When Cavanaugh asked Moscow Centre to explain why it did not report the event to WANO, Toukhvetov agreed to send the information to the Managing Director. Carlier said that the centre's actions were "unacceptable", but the Governing Board would not push the issue. For Cavanaugh, the whole affair heightened the importance of and necessity for greater accountability of regional centres and governing boards – as well as the dangers of failing to report and share operating experience among WANO members.[64]

Kingsley, who attended the Edinburgh meeting and was furious at the response, added accountability for the Special Committee to include in its recommendations. Over the next nine months, the Committee collected information, met with past and current WANO leaders and questioned WANO members, especially utility executives and regional governors and directors. Not surprisingly, in Asia where much of the new plant construction was occurring, governors and executives requested more WANO aid to support the expansion. SK Jain of India wrote that WANO's Operating Experience and Peer Review programmes were "high-quality programmes" and in the future WANO should "deploy some of its resources to address the needs of plants under construction and also ageing related issues in older plants". The Tokyo Centre representative, Jianfeng Yu of the China National Nuclear Corporation, suggested adding "improving plant performance" to WANO's mission as well as a programme tailored for plants under construction. He also saw the need for training a larger regional staff for Tokyo Centre with the increased number of plants expected to begin operation in the future.[65]

Kingsley reported the results of the Special Committee's work at the Governing Board meeting in Québec City in October. The major recommendations were no surprise: more accountability at all levels; greater CEO involvement; better training for peer reviewers; more frequent peer reviews and follow-up; longer staffing assignments to regional centres; strengthened WANO central governance; more responsible and responsive regional governing boards; improved plant performance; and shifting accountability from the centre directors to the regional board chairmen. The report, the Governing Board determined, would be folded into the WANO Long-Term Plan. One reviewer commented that the Special Committee seemed to be leaning toward central governance and that WANO regional centres should be more like INPO. Kingsley agreed. "A worldwide WANO is too big, and strong INPO-like centres are needed."[66]

The report of the Special Committee ended in a whimper. Placing the recommendations into the WANO Long-Term Plan defused any urgency and action. In addition, the resistance of regional governors toward assuming accountability to drive plant improvement disappointed Kingsley and his allies. Only the idea to hold special meetings for CEOs to encourage their participation in WANO was implemented. With just six months left in his presidential term, Kingsley turned his attention to the ninth BGM, to be held in Chicago in September 2007, to ensure its success.[67]

And by all accounts, a success it was. The closed meeting for CEOs, based loosely on an INPO model, was well attended, although it was not as hard-hitting as Kingsley and other INPO veterans had wished. Others thought that the tone of the meeting was inappropriate for an international group; the room was too large to engender discussion among the CEOs, many of whom were meeting together for the first time. The meeting became a one-sided conversation where Cavanaugh and Mampaey did the talking. Nonetheless, many recognised the potential of further engaging CEOs and future sessions were planned. Although the Special Committee's recommendations were not fully embraced by WANO and regional governors, the logic behind them took centre stage at the BGM. Cavanaugh hammered home the lesson of collective responsibility. The future of nuclear energy, he told the delegates, "depends on the safety of your own fleet and that of your colleagues. I submit that some have forgotten – or do not understand – their collective responsibility for working together to improve performance. WANO is a voluntary organisation, but in the apparent view of some members today, membership does not include any serious sense of obligation." Membership came with "a limited number of reasonable obligations, including prompt reporting of events and use of operating experience, the use of peer reviews as effective tools for improvement", and to "strengthen WANO's infrastructure by providing quality resources and leadership". If the Governors would bottle up the Special Committee report, Cavanaugh could at least let the genie out into the open.[68]

****

For all the planning to improve WANO's future, the quandary of succession still lurked over the association's best intentions. As his term drew to a close, Cavanaugh outlined to the Governing Board what could be expected in 2008. Since beginning his second term, the WANO Governing Board had changed dramatically. Stane Rožman had replaced Stricker as Chairman of the Paris Centre Governing Board, Duncan Hawthorne had replaced Hairston as Chair of the Atlanta Centre Governing

Board, and James O Ellis, the CEO of INPO, had succeeded Gary Gates, a longtime representative from Atlanta Centre. Two other new governors from Paris and Tokyo also joined the board, but one from Ukraine showed little interest and was soon replaced; the other, from Pakistan, had continuing visa issues and attended meetings infrequently during his term. In 2005, Takashi Shoji joined Tokyo Centre as Deputy Director, later becoming Director. There were also personnel changes at the regional centres. In late 2005 Dave Igyarto had assumed the position of Director of Atlanta Centre from Ed Hux, and the next year Ignacio Araluce replaced Canaff as Director in Paris. In 2007 Mikhail Chudakov replaced Toukhvetov in Moscow. In addition to Cavanaugh, Mampaey was scheduled to leave in 2008. With so many recent changes to WANO's governing structure, selecting a new chairman and managing director took on special urgency.

Then, at the end of November 2008, a group of terrorists attacked the Indian city of Mumbai, horrifying the world as it watched events unfold on television over the next several days. The WANO Governing Board meeting scheduled to be held in New Delhi two weeks later in anticipation of the tenth BGM, to be held in that city, was postponed. With security concerns at a peak level, some governors thought many members would not attend a BGM in India and that an alternative site must be found. But Dr Jain assured Cavanaugh and most of the Governors that India would return to normality and that security would be tight. Jain showed the board a letter from the Indian home secretary, taking personal responsibility for the safety of anyone attending the BGM. WANO would meet in New Delhi, but the date was postponed until early 2010 to assuage security concerns and assure a healthy registration.[69]

Change and uncertainty, the governors realised, could cause very rough waters. Moreover, altering the course and part of the cultures of an international organisation would take time. But after 2002 it was clear that the course had been changed.

However, few of the proposed changes had been implemented. The causes for WANO's inaction were many: the failure to create an effective and united Executive Leadership Team; the lack of clout of the managing director, a position that carried more responsibility than authority; the friction between the main Governing Board and the regional boards' autonomy; the failure to change WANO fundamental governance documents to match new policy direction; the drop in the direct involvement of top nuclear utility executives in WANO; and the turnover and lack of continuity of both WANO and regional governors all contributed to the failure to act decisively. How the next chairman steered the WANO ship through the fog of uncertainty ahead would reshape WANO's future.

Chapter Six

LAST CHANCE **TO GET IT RIGHT**

In the 20 years since the founding of WANO in 1989, the international nuclear power industry sought to prevent another Chernobyl through continuous safety improvements among the world's nuclear utilities. Obstacles that the industry needed to overcome were significant – sharp differences in cultural norms, linguistic and communication issues, lack of adequate financial resources, a shortage of trained professional staff, complacency born of an attitude that shoddy safety practices happened elsewhere and an unwillingness to report problems or share critical information. In addition, government ownership or control over utilities in some countries reduced WANO's leverage. Nevertheless, WANO's programmes had guided steady improvements across the global fleet of nuclear reactors. Those successes, however, also led to a declining sense of urgency as memories of Three Mile Island and Chernobyl faded. For some nuclear operators, commitment to WANO's goals slipped from active participation to pledges and promises. Lulled by success, some operators saw little value in WANO's insistence to achieve excellence in nuclear power operations.

WANO was trapped in a tyranny of time. Appearance was allowed to replace reality. As the period between serious nuclear accidents lengthened, some in the industry – too many, according to experienced WANO hands – grew complacent, lulled by a sense that they had achieved a strong safety culture and that WANO's programmes, services and warnings were therefore less critical. Rosenergoatom, which managed all the Russian nuclear plants, exemplified the problem. "Currently," a Moscow Centre representative reported in 2008, "Rosenergoatom places little emphasis on WANO based on the belief that the performance of the stations is 'quite acceptable'." WANO veterans knew better; they believed that the declining participation or lack of engagement of top utility executives in WANO was symptomatic of the failure to understand an exceptional aspect of nuclear power – that the operation of one plant would impact the future of all.[1]

By 2009 Pierre Carlier worried that so much time had passed since the "last big event" that overconfidence and complacency had become widespread among nuclear operators. In his mind, 20 years of success did not translate into ensuring operational safety. WANO's core programmes worked well when the members worked together, when they remembered that "we are all hostages of one another". But all members did not fully participate. Differences among the regional centres further undermined the organisation's goals. A number of events had occurred that could have resulted in serious consequences, so there was still plenty of room for operational improvements at existing plants and many lessons to be learned by new entrants to the nuclear power community. A retreat from WANO's programmes and mission, Carlier maintained, left WANO's members "dancing on a volcano".[2]

Carlier's volcano metaphor was apt. The danger of a potential destructive event was ever present, no matter how well the industry believed it was performing. His concern drew on an axiom of a fellow countryman more than two centuries before. Voltaire had said that "best is the enemy of the good." The difficulty lay in setting standards that constantly called for improvement, to strive for perfection knowing that it can rarely, if ever, be attained. Nuclear plant operators could not relax safety standards; accident prevention must remain paramount. Because the risk was constant, the response had to be uniform and unrelenting, invariably vigilant, Carlier and other WANO officials believed. To do otherwise was to gamble with safety.

****

The man who had inherited WANO's rumbling volcano was its new Chairman, Laurent Stricker, a protégé of Rémy Carle and Carlier at the French utility, EDF. A native of Nancy, a city in northeastern France near the German border, Stricker graduated from the Polytechnique Grenoble and l'Institut National des Sciences et Techniques Nucléaires de Saclay and, after a brief career as an instructor for a government commission, joined EDF as a young engineer in the late 1970s. Over the next 35 years, Stricker rose to become Head of Nuclear Operations, responsible for the operation of the entire French nuclear fleet of 58 nuclear units, and deputy general manager of EDF's Generation Division. When he retired in 2007, he became special nuclear adviser to the President and CEO of EDF. Carle had invited him to attend the Paris BGM in 1995, but little resulted from the experience. EDF's operators, including Stricker, were convinced at the time that "we didn't need others to tell us how to do our job correctly". A few years later, Carlier, then head of EDF's nuclear division and increasingly uncomfortable with his company's isolated views on safety, pushed Stricker to become more involved. He explained that WANO "was very important to open our eyes and our minds outside of our own". Still, Stricker held the view that WANO was "not very important in general to improve our organisation".[3]

In 2003 Stricker's involvement with WANO took on a new dimension when he became a WANO Governor and Chair of the Paris Centre Governing Board. "I started to see at this time how WANO was working," he later recalled. "To be completely candid, I wasn't very impressed." Stricker was not convinced that the structural change from a WANO coordinating director to a managing director had improved the organisation. Issues between the WANO Governing Board and the regional boards remained difficult. "The decisions[s] of the main governing board [were] not clearly implemented within the regional boards," Stricker observed, in large part faulting the friction between Berg and the regional directors for the failure. The centre directors did not accept Berg's new authority or his "strong personality". Berg was inclined to make decisions and stick to them rather than include others in decision-making, convince them of the policy's necessity or discuss disagreements in an effort to reach consensus. "If a decision was not accepted," Stricker said, the directors made "no attempt to implement it". For the governing boards of Paris and Moscow, the view was "we have to take care of the regional business ourselves. It is not the role of London. That made it very difficult to work closely together. I was convinced that I was able to manage the regional centre better than the Managing Director."[4]

Nevertheless, Stricker saw the value in a more centralised WANO. He appreciated what Paris Centre director Yves Canaff had accomplished but admitted that "this guy wanted to work alone." Yet he recognised that "cooperation between the four centres was very, very, very poor. I think it was a mistake not to cooperate completely." For the two years that Stricker chaired the Paris Centre Governing Board, he remained cool on the effort to consolidate authority in a managing director and a dogged defender of regional autonomy and the strength of EDF's operations. But he also recognised WANO's strengths. It could be valuable for nuclear operators, and EDF was every bit an exemplar. In Stricker's mind, the time was right to elect a chairman from Paris Centre. As Cavanaugh's term came to an end, Stricker advanced his own candidacy as a possible successor.[5]

For two years Cavanaugh and the Governing Board had canvassed regional board members and member CEOs to provide names of potential candidates for the post of WANO chairman. In the summer of 2007 he named W Gary Gates, the president and CEO of the Omaha Public Power District and a WANO governor from Atlanta Centre, and Stane Rožman of Paris Centre as the Nominations Committee. The men were to make their final recommendation at the Governing Board meeting in Helsinki the following April. In February, Cavanaugh, Rožman and Mampaey interviewed three finalists, all Europeans from Paris Centre, in London. The committee did not recommend Stricker. Rather, it selected Dr Walter Hohlefelder, who was retiring from the management board of the large German utility, E.ON Energie AG. Hohlefelder also served as President of the German Atomic Forum. Like Maeda, Hohlefelder had limited WANO experience, and some governors thought his election should be delayed until a Special Committee of regional governing board chairmen adopted or rejected some proposed modifications in the WANO governing structure. Part of the motivation was to ensure that there would be strong direction and leadership from London while a new chairman became more familiar with WANO.[6]

The election and the changes were clearly interrelated because, if adopted, they would have considerable impact on the powers of the new chairman. After lengthy discussion, the WANO governors nevertheless decided to move forward with Hohlefelder's nomination, "agreeing that the chairman-elect be fully informed of the proposed actions before accepting the position". The motion to elect the new chairman also contained a proviso to make Cavanaugh "a special adviser to provide a complete handover and operational support for the new chairman". While a majority of the centres approved the motion, great uncertainty and tension remained. The proposed changes had not been fully approved, the candidate was unaware of what was in the wind, and not all governors were comfortable with the process or the result.[7]

Uncertainty in the nomination procedure stemmed from a series of proposed modifications to the WANO Articles of Association and Charter. The first, to incorporate the role of managing director as part of official documents, was uncontroversial.

However, a second, to establish a permanent chief executive officer position in WANO, generated lengthy discussion. The CEO proposal would "allow for more continuity of leadership in light of the two-year terms of the WANO Governing Board chairman and the managing director", Cavanaugh explained. Many of the Governors saw a need for such a position but only if the roles and responsibilities in relation to the chairman and regional centre directors were clearly established. Moreover, another layer of bureaucracy would raise personnel costs in London – something the Governing Board watched warily.[8]

The governors were uneasy with the title of CEO and what authority it might hold, and the Governing Board suggested "director-general" as an alternative. Another director asked the Governing Board to consider adding a third governor from each region and establishing a vice-chair position for the Governing Board. Most backed the idea of an additional governor as it "would provide more stability, engage more members, and allow more flexibility in attendance at meetings". To provide continuity in the event of a transition to the proposed restructuring, the governors approved extending Mampaey's contract for an additional year, into 2009, and adding a second deputy director to assist with the additional work anticipated with the expanded Governing Board.[9]

Shortly thereafter, the plans unravelled. After speaking with Cavanaugh and Mampaey about details of the WANO chairmanship in May, Hohlefelder withdrew his candidacy, outraged that the proposed changes in governance would be made without his advice or consent. The governors, meeting in a series of telephone conferences and emails in June, voted that Cavanaugh continue his term until August 2010 or until a replacement was found. But the vote was not unanimous; some governors, uneasy with the extension and Cavanaugh's manoeuvring, asked that a new Nominations Committee be appointed immediately to find a successor.

Furthermore, several governors questioned the urgency of the proposed changes, particularly the idea of a director-general, whose salary they could not approve "when there is such a strong need for resources in the regional centres". Once a new chairman had been elected, that person could determine "when (or if) further action" on the proposals was warranted, the opponents to the changes argued. Once again, the search for a successor had not gone smoothly.[10]

Stricker's opposition to the reduction of regional autonomy and his backing of Canaff against Berg were part of the reluctance of the Governing Board's Nominations Committee to put his name at the top of the candidate list. Stricker was widely perceived as an EDF, not a WANO, man. But with Hohlefelder's withdrawal, Stricker once again became a candidate. Carlier, who had grown impatient with Cavanaugh and his inability to create a consensus for strengthening WANO, urged the top executives at EDF to support Stricker as well as WANO's Peer Review programme, in which the French utility had only grudgingly participated. In addition, Carlier urged Stricker to create an agenda for WANO, strengthen the WANO Governing Board, and get CEOs more deeply engaged in WANO activities. Stricker, who had learned from the head of EDF that the meeting of CEOs at the Chicago BGM in 2007 had accomplished little but held promise, embraced CEO involvement as his main objective. In January 2009 Stricker was elected chairman of WANO. With his election, the push for the more radical changes in WANO governance that may have caused Hohlefelder to withdraw ended, but Stricker supported the other modifications pushed by Cavanaugh and the WANO governors to expand the Governing Board and include more CEOs.[11]

As Stricker took over the chairmanship of WANO, its Managing Director, Mampaey, was preparing to leave. To follow WANO's pattern, the successful candidate to replace him would be a native English speaker. When James O Ellis, the CEO and President of INPO, asked his colleagues for suggestions on who might fill the position, George Felgate, a longtime INPO employee with substantial international experience, said he would like to be considered for the job. Felgate was a true Yankee, born and raised in Connecticut. He graduated from the US Naval Academy in 1970 and went into the nuclear propulsion programme, serving on two submarines in his 10-year career, with time ashore to complete a master's degree in computer systems management at the naval post-graduate school in Monterey, California. After leaving the navy, he worked at the Fast Flux Test Facility (FFTF) in Richland, Washington. In 1982 he joined INPO, where he caught the eye of Zack Pate. In 1989 Pate recommended Felgate to his friend Kenneth M Carr, a new appointee to the Nuclear Regulatory Commission and a former admiral and submariner. Subsequently, Felgate served as Carr's executive assistant for two years, managing all office activities and advising the new commissioner on nuclear power operational and maintenance matters that came before the Commission.[12]

After two years with Carr, Felgate returned to INPO and worked in various areas such as training, emergency preparedness, operations and human resources. He soon rose to Vice President. In 2000, at the urging of Sig Berg, he took a job as Operations Manager at the Koeberg Nuclear Station in South Africa and then returned to INPO. In 2008 he was the team leader of the first corporate peer review ever conducted in Japan of TEPCO. There he met Stricker, who was a senior adviser to the peer review team. He liked Stricker and thought they "hit it off". That meeting served as a background for Felgate's interest in the managing director position. Felgate had a reputation as a detail man and a tireless worker. His personal attributes and experience hit the mark with Stricker. In August 2009, Felgate was on his way to London. The two men proved to be a compatible and effective team, travelling together carrying the WANO message for change.[13]

****

Like many international organisations, WANO consisted of a group of allies seeking similar goals, often with different sets of ideas, assumptions and rules. The shared paramount goal of WANO's members was the safe and reliable operation of nuclear power plants. To achieve that, WANO had established the four core programmes of (1) operating experience to share critical information and learn from the experience of others; (2) professional and technical development to improve professional knowledge and skills; (3) technical support and exchange to resolve problems and improve safety; and (4) peer reviews to enable plants to measure themselves against the best practices worldwide. If frequently used and properly employed, the core programmes could make a tangible difference in plant safety and reliability, the WANO leadership believed. The key element was how the programmes would function under a joint operation with a chairman and a central governing board to set policy, four autonomous regional governing boards, each led by an independent director, and the coordinating centre to administer policy. However, no single entity had the authority to ensure or enforce the implementation of WANO policy. Only once previously had WANO demonstrated its clout after a peer review found serious problems at the Chernobyl plant in 1996. At that time Pate had assembled considerable international pressure on Ukraine to fix the problems. But memory of that success had faded. While the organisation could reach a policy consensus, it could not readily determine how that policy might be implemented. By 2009, WANO could offer many carrots, but it held few sticks.

Accordingly, WANO, as it gained more experience over the years, sought to change – to centralise its governance and have more authority, accountability and enforcement. The creation of the position of managing director in 2002 was the first step in WANO's "journey of continual improvement". For the better part of a decade, the WANO Governing Board had laid the foundation for fundamentally altering the organisation's mission statement and its governance and membership structures, thereby positioning "ourselves to better navigate an ever-changing nuclear landscape". The updates reflected the combined feedback of industry leaders who had shared their views with WANO during a series of reviews and sponsored forums over the years. Stricker, as Chairman, now backed these plans. With Felgate, Stricker held a series of face-to-face listen-and-learn meetings with CEOs, explaining the need for change and gathering suggestions for the shape these changes should take. The encounters always included an appeal for more direct CEO involvement in WANO activities. In addition, at the urging of Stricker, former CEO of British Energy William A Coley conducted a series of "highly productive" small-group CEO meetings that contributed to WANO's blueprint for change. Still, there was much work to be done to bring that blueprint into reality.[14]

The nuclear power landscape had shifted significantly since WANO's founding 20 years before. On the positive side, through its Peer Review programme, WANO had become recognised as one of the major international nuclear safety organisations. Multinational companies owned or operated plants in a variety of countries, and many nations were served by multiple corporations. A multinational WANO had helped lead the way. John Ritch, Director-General of the World Nuclear Association, which served as a forum for nuclear energy and had partnered with WANO to establish the World Nuclear University, noted that 20 years after Chernobyl, WANO's programmes represented "nothing less than a foundation stone on which our entire industry stands". The promise of a nuclear renaissance was at hand, many believed. There were more than 55 units under construction by the end of the first decade of the 21st century and many more in the advanced stages of planning. Moreover, many environmentalists had adopted nuclear energy as a crucial part of the battle against carbon emissions and global climate change. Surely, the future was bright.[15]

In addition, WANO's programmes were performing well, and the metric of the nuclear industry had improved in the two decades since the organisation was founded. Improvements in nuclear safety and reliability were particularly important as plants aged and new players and new designs emerged. Importantly, the decline in accidents occurred even as the number of units reporting data rose from 169 to 216. There was also an improvement in operational performance and fewer emergency shutdowns; the rate of scrams declined from 1.8 per 7,000 hours of critical operation to 0.4 over the same period. The average unit capability, a gauge of a plant's cost-effective reliability and generation output, climbed from 77.2% in 1990 to 87.3% in 2009. The industrial safety accident rate, which tracked the number of accidents that resulted in lost work time, restricted work, or fatalities per 200,000 man hours, fell from 5.2 to .78 between 1990 and 2009, a drop of about 85%, according to WANO statistics. Ironically, these positive performance statistics were exactly what worried WANO veterans – that a generation of executives and operators had little appreciation for the significance of a Chernobyl or a Three Mile Island.[16]

To Carlier, the loss of the lessons of history triggered visions of the volcano rumbling below. WANO still operated on the edge. In spite of the favourable metrics, WANO's road to excellence was pitted with potholes – a decline in CEO involvement, a growing complacency among nuclear operators that all was well, and the fact that the schedule to conduct peer reviews at each nuclear unit at least once every six years, rather than four, was a recognition of reality rather than a requirement of safety. Further, the continued reluctance of members to share operating experience or report events belied WANO's underlying principle that accidents could be prevented if lessons were learned from previous incidents. A 2009 survey of peer review summaries revealed that "global trends include shortfalls in setting and reinforcing standards by managers and inappropriate supervisor and worker behaviour" and an "ineffective use of error prevention techniques". The themes or areas for improvement were not new and indicated that progress "had been less than desired". Change, WANO leadership maintained, "walks hand-in-hand with progress". One could not occur without the other. "The structure and governance of WANO that was so necessary when it was formed," Stricker said, "limits us today". As a result, WANO experienced more change in 2010 than in any year since its creation.[17]

****

Since his election as WANO Chairman, Stricker had shed his regional perspective for a more centralised organisation. The impetus for many of the changes at WANO came from Stricker, who became WANO's premier salesman. He spoke at nuclear conferences, gave press interviews and travelled widely to meet with CEOs of nuclear energy companies. Over the course of two years, he and other WANO officials led by Coley held a series of meetings in Paris and London with small numbers of utility CEOs, 10 to 12 at a time, eventually reaching a large percentage of the nuclear industry in this manner. The intent was to get CEO input and discussion regarding where WANO needed to move in the future. The small numbers of participants encouraged candid discussions, emphasised the value of WANO programmes and served to re-engage the utility executives in WANO activities. Coley and Stricker discussed the value and importance of WANO in improving the safety of their nuclear power plants and stressed the importance of their personal involvement. In describing WANO, Stricker drew on an automobile analogy. What WANO did was "similar to taking care of the car driver and the upkeep of the car, but not taking care of the type of car selected". In other words, it was the safety of operating the vehicle that was crucial to WANO, not the make or model of the car. WANO would do everything in its power to make sure a plant was safely operated and maintained. It was the responsibility of the driver "to ensure that the right type of fuel is used, the maintenance is done regularly and the car is in top operating shape". The information gathered at these meetings would play an important role in drafting the governance changes presented to the membership at the New Delhi BGM in 2010.[18]

Stricker pushed the benefits of peer reviews for both plants and corporations. They were "a win-win for WANO and the utilities involved", he explained. He stressed the importance of the utilities and WANO working together to upgrade poorly performing plants. At Stricker's urging, the sessions also elicited suggestions from the CEOs on how WANO's role and activities could be improved. He and Felgate travelled extensively together as a team, meeting with CEOs, the Chairman emphasising the collective responsibility between CEOs and WANO for nuclear safety and the Managing Director describing how WANO could help. From these meetings and forums a number of recommendations emerged that would lead to rethinking WANO's structure and operations. These recommendations became the basis for the reforms presented at the New Delhi BGM.[19]

With the 10th BGM occurring just months after the two took on their new WANO roles, both Stricker and Felgate had to hit the ground running. As Stricker focused on CEOs, Felgate turned his efforts to making the Executive Leadership Team an effective body. Broken under Berg and somewhat improved under Mampaey, the ELT could still be fractious. It was Felgate's good fortune that most of the regional directors changed in 2009 and the ELT's good fortune to have a managing director committed to making the ELT an effective group to implement WANO programmes. "I approached our work in a spirit of compromise," Felgate said later. "We would often cut 'deals' with a director who had dug his heels in on an issue. If he would support the ELT on issue 'X,' we would back off on initiative 'Y.' In doing so we were able to move forward on most initiatives, while sacrificing a few." One example concerned WANO's programme to woo CEOs. Felgate and the staff in London developed an annual letter from Stricker to each member CEO. The letters, Felgate recalled, "were blunt and hard-hitting". Each described the performance of the units for which the CEO was responsible, how the performance indicators stacked up for that utility, and, most importantly, "how that CEO was supporting WANO based on his personal participation and how many peers he allowed to participate in reviews and technical missions". While some regional directors were uncomfortable with this initiative, Felgate, backed by the majority of directors, earned the backing of the entire ELT. Such a collegial team would be required to achieve approval of the momentous changes proposed in India.[20]

****

The theme of the 10th BGM, held in New Delhi in February 2010 and hosted by the Nuclear Power Corporation of India Limited (NPCIL) and Tokyo Centre, was "Moving Forward Safely – In a Changing World." The theme reflected WANO's strategy to keep pace with a series of changes in the industry. It was a landmark meeting. "WANO is at a crossroads," Stricker told nearly 400 delegates, explaining that the organisation had charted a new course to help navigate "an ever-changing nuclear landscape". What had been successful in the past, he said, "limits us today. We need a fresh look at WANO. WANO needs to adapt." Convinced by Stricker's words, the delegates agreed. In an ornate ballroom in the Taj Palace Hotel, they voted unanimously to support the first significant changes in membership and governance since WANO's founding in 1989. In another notable change, the delegates elected Qian Zhimin, Chairman of the China Guangdong Nuclear Power Group (CGNPG), as president of WANO. The CGNPG was the largest company in the world in terms of nuclear power generating capacity under construction, a recognition of the importance of new plant start-ups and WANO's critical role in the safety of the nuclear industry. Moreover, the next WANO BGM would be held in Shenzhen, China, in October 2011.[21]

WANO's first step was revising its original 1989 mission statement to place greater emphasis on working together to achieve better performance. The organisation's scope of activities had expanded over the years, and the updated mission statement reflected this. WANO's revised goal was "to maximise the safety and reliability of nuclear power plants worldwide by working together to assess, benchmark and improve performance through mutual support, exchange of information and emulation of best practices." Although the statement did not specifically mention peer reviews, the inclusion of assessments indicated that over the years, peer reviews stood "at the very heart of WANO's programmes" and had been accepted by most members as "a catalyst to improve plant performance".[22]

The governance provisions put forward at New Delhi were the most sweeping organisational changes since WANO's founding more than two decades before, and approval, according to one participant, was "no slam dunk". The membership structure was completely overhauled. The organisation's mission shifted, as did the composition of the Governing Board. Expansion of membership and voting rights were significant changes. For WANO's first 20 years, only one member, representing all the utilities operating in a given country or region, was able to cast a vote. "That was the right approach at the time," Felgate told *World Nuclear News*, "but times have changed." Over the years, consolidations and mergers had resulted in large utilities operating plants in many countries. In this new environment, WANO needed more direct involvement by the CEOs who set the priorities and direction of these multinational corporations and, importantly, controlled the purse strings and thereby had a direct impact on nuclear safety. Success in a region was dependent on what Felgate termed the "big dogs", the most "powerful influential member in each region who could control voting at the Regional Governing Board level and could keep the smaller utilities in line". INPO, EDF, Rosenergoatom and TEPCO were all big regional players. Changes to the membership structure would "improve member engagement and increase resources dedicated to WANO programmes". The WANO membership was realigned to reflect operating companies as the primary members and encourage more operating companies to become members with voting rights. Under the revised system, each voting member received five votes plus one vote for each nuclear unit they operated or represented. The change gave more input into WANO's activities to those entities commensurate with membership fees and services provided, "while ensuring smaller players are fairly represented and continue to hold voting rights". Felgate concluded that "had we not had an effective working ELT, none of this could have been accomplished."[23]

Membership of the WANO Governing Board was also expanded. Frustrated by failing at times to achieve a quorum, the Governing Board initially allowed alternates to replace a missing board member. But for a long-term solution, the Governing Board recommended an expansion to three members from each regional centre and the WANO President, increasing the Governing Board from nine members to 14. Involving more CEOs in WANO's governance was behind the revision. "It is expected that board members will be member CEOs or very senior executives," WANO explained to its membership. The regional governing board chairman would automatically be a member of the WANO Governing Board, and regional boards would nominate two additional candidates from their region, "one of who is expected to be the CEO who represents the region's largest or most influential member." In addition, the WANO president, a largely honourary position with the responsibility for planning the next BGM, became a voting member of the restructured Governing Board and – it was hoped – an active emissary for WANO. The bottom line was that WANO members would be "better represented by high-level decision makers within the regions." The proposal was adopted at the New Delhi BGM and the offices filled during an Extraordinary General Meeting in Murmansk, Russia, in July 2010.[24]

With the revision in membership came revisions in the fee structure as well. WANO centres, with the exception of Atlanta, had always encountered a shortage of resources, be it in funding or staffing. The new arrangement included WANO reimbursement of expenses for secondees provided by members, whereby a portion of membership fees were paid to London and a regional affiliation fee paid to the regional centres. In addition, the name of the Coordinating Centre was changed to the London Office, recognising the functional changes that had occurred since the establishment of the managing director.[25]

Finally, the sweeping changes approved in New Delhi strengthened members' obligations and commitments to WANO and redefined the roles of the managing director, London Office staff and regional centres. In order to adjust the membership to the revised definitions, members and prospective members were asked to reapply for WANO membership in April. The revised membership was approved by the Governing Board in June, resulting in a total of 94 members in three membership categories. By the end of 2010, WANO had 97 members, including the Emirates Nuclear Energy Corporation, the first new entrant from a country without an existing nuclear programme.[26]

The changes to WANO's governance structure had been fast in coming, but approval had not been assured until just days before the BGM. "The outcome was in doubt up to the last minute," Felgate recalled. "We heard that the CEOs from Moscow Centre were not going to approve the changes. Laurent [Stricker] was on the plane to Moscow days before the meeting urging their support for the good of WANO. The magnitude of the changes was immense," he stated. WANO had changed its mission, enlarged the composition of its Governing Board, overhauled the membership and fee structure and put in place details to move from a Coordinating Centre to a London Office. Stricker's pleas worked. With Rosenergoatom leading the way, Moscow

Centre backed the changes.[27]

Implementation, however, encountered significant delays. While the impediments to change varied by region, adapting to the new fee structure and meeting secondee expectations were "the most difficult issues", WANO admitted in its 2010 Annual Report. While the London staff increased from 11 to 18, some regional centres, particularly Moscow and Tokyo, needed to stretch out their staffing plans over three to five years. The plans put forth by the two centres did not appear to provide sufficient resources "to maintain the current level of support, much less improve, with the rate of new entrants into the field" and new plants currently scheduled to begin operation. "Realistically," the Report dryly noted, "this will slow the anticipated progress."[28]

The tough tone of the 2010 Annual Report reflected the changes in governance and accountability that had been introduced in New Delhi. With a limited distribution among members, the review made clear which regional centres were not toeing the mark set in the revised Charter. But the Report also clarified the expectations of the association. The Report focused on four WANO principles – a member's individual responsibility for nuclear safety; members' collective responsibility for nuclear safety; WANO's governance, staffing and resources; and WANO's visibility with its stakeholders. Each member had an obligation to ensure that its units are "operated to the highest standards of nuclear safety and reliability" using frequent peer reviews, technical support missions, corporate reviews and the timely implementation of WANO's Significant Operating Event Report (SOER) recommendations.[29]

In addition to a member's individual responsibility, WANO emphasised each member's collective responsibility to ensure that "every other nuclear station in the world is operated to the same high standards" of safety and reliability. This was part of WANO's founding principle that "we are all hostages of one another" and made the quality, quantity and timeliness of reporting operating experience crucial to preventing, or at least limiting, similar events elsewhere. Performance indicator information would promote benchmarking and assist plants in monitoring and measuring performance results against other units – though many stations, which the report identified, did not report complete data. Regional governors, the report maintained, needed to be more involved in peer reviews, technical support missions and WANO-sponsored seminars and workshops. Changes in governance were part of the effort to promote a stronger WANO – with both sufficient financial and staffing resources – in order to meet the association's obligations to members and ensure its longevity.[30]

The last area of the Annual Report's focus was WANO's visibility. The idea of WANO projecting a higher profile in the world nuclear community had been discussed for many years, usually pushed by Moscow Centre. But the concept failed to gain much traction among the WANO governors, who worried about the confidentiality of plant-specific data and concentrated instead on private communications issues among the members. Translations of WANO products such as SOERs and other technical reports, as well as encouraging access to WANO's website for operating experience and performance indicator information, represented lengthy efforts to improve communications among plant operators. The publication of the quarterly *Inside WANO*, which Carlier instituted in 1993, was targeted to members to assist with their understanding of WANO's vision and direction, and also provided features on the nuclear plants and people involved with WANO activities. Yet as late as 2009, Felgate recalled going to an international conference on nuclear safety and never hearing WANO mentioned. "I went away from the conference shocked [and] disappointed, but committed to turn that perception around." With Stricker's support, WANO drew up a new communications plan to raise WANO's prominence.[31]

LAST CHANCE **TO GET IT RIGHT**

As the nuclear power landscape changed, WANO responded by including a wider pool of stakeholders to whom it would distribute its information. "Communication to key stakeholders about who WANO is and what WANO does is becoming increasingly important," the association stated, and that circle now included the public, the media, regulatory authorities and other nuclear industry organisations. The 2010 Annual Report came out in January 2011. While it outlined the successes, the remaining gaps for improvement, and the promise of the New Delhi reforms, it could not have anticipated the impact of Fukushima.[32]

****

On 11 March 2011, a massive earthquake followed by a towering tsunami slammed into Japan's east coast, overwhelming the nuclear units at the Fukushima Daiichi nuclear power plant operated by TEPCO, the country's largest utility. It was the perfect storm. While the plant was designed to withstand a powerful earthquake, and sea defences were in place to protect the station from tsunamis, no one had foreseen the level of devastation that resulted when the two occurred together at unprecedented levels.[33]

The earthquake was the most powerful ever to hit Japan, 9.0 magnitude on the Richter scale, moving the main island of Honshu some two-and-a-half metres. The resulting tsunami, one of the most destructive on record, sent waves as high as 14m crashing ashore. Although the boiling water reactors suffered little earthquake damage and shut down automatically as designed, the flooding caused by the tsunami disabled all emergency power sources and led to a series of equipment failures, multiple hydrogen explosions, three melted nuclear cores and significant releases of radiation in civilian areas beyond the plant site. The earthquake and tsunami shook, then quickly eroded, the world's confidence in nuclear power. After Chernobyl and more than two decades of work to improve nuclear power safety, WANO members learned once again that the industry was only as strong as its weakest link.[34]

Twenty-five years after Chernobyl, through nearly a quarter of a century without a major nuclear accident, Carlier's volcano had erupted, though in a manner he could not have imagined. The aftershocks of Fukushima were global. The nuclear industry was stunned. Japanese policy-makers questioned TEPCO's safety culture and its response to the catastrophe, suggesting that the country might abandon nuclear power altogether. Government officials in Germany, Switzerland and Italy sought to end or phase out nuclear power, as did governors in New York and Vermont in the US. Talk of a nuclear renaissance fell silent.[35]

WANO staff followed the events of March 2011 in real time. A member of WANO's London staff, Takashi Shoji, was on the phone to Tokyo Centre when someone on the other end of the line said, "Oh my God, we're having a big earthquake." Felgate contacted his former colleagues at INPO in Atlanta, urging them to man its emergency response centre, as WANO had nothing similar. As an organisation that had been created to promote accident prevention and safe operations, WANO had no experience in accident mitigation. WANO had no emergency plan. Stricker and Felgate decided that WANO should keep members informed of what was happening in Japan and offer the country and TEPCO the assistance of all the world's nuclear utilities. The London Office became a clearing house for requests for boron and consumable items such as disposable gloves and respirators needed in the emergency. London sent on requests to member utilities that might have the materials on hand to send to Japan. But "that's when we ran into headaches," Felgate recalled. "We would have the boron identified, the plane identified and then we couldn't get clearance [from the Japanese government] to get it shipped into Japan."[36]

Without an emergency plan or significant contacts with vendors and suppliers, WANO could provide little leadership on its own and soon found itself in a secondary role. INPO, on the other hand, under President and CEO James Ellis, could react immediately. It staffed its emergency response centre around the clock, sent a team to Tokyo to work with TEPCO and opened lines of communication with the Nuclear Regulatory Commission and other US government agencies, where Ellis had many senior contacts. Even so, INPO, too, had difficulties establishing open communications and trust with TEPCO, but it pushed ahead, organising and leading an industry support team consisting of staff from the INPO and the US Electric Power Research Institute, suppliers and industry executives to work with TEPCO. INPO teams remained in Japan for months after the accident, creating a strong bond of partnership and mutual trust among INPO, TEPCO and other Japanese nuclear organisations, such as the Japan Nuclear Technology Institute (JANTI).[37]

Though always WANO's strongest supporter, INPO, through its actions, effectively strained its relationship with WANO. As a member of WANO, INPO could have said: "We'll do this on behalf of WANO," Felgate said later, but the response was not done under either the WANO or Atlanta Centre banner. Clearly, INPO had the resources and the expertise, and WANO did not. But the effect of INPO's strong and immediate response, some WANO officials thought, was to somewhat undermine the credibility of WANO.[38]

INPO's credibility was riding high. After the disastrous well blowout and oil spill at the Deepwater Horizon drilling platform in the Gulf of Mexico in the summer of 2010, INPO had been singled out by the Commission investigating the accident as an effective, independent, self-policing model for correcting operational problems in high-tech industries. Ellis and Pate testified in nationwide televised hearings about INPO's history and its positive impact on the nuclear power industry in the US and abroad. For most Americans watching, the hearings were the first time they had ever heard of INPO and what the institute did. In contrast with the muddied waters of the troubled oil industry, INPO presented a positive beacon as a vigorous and vigilant champion of safety. The Deepwater Horizon Commission's report praised the INPO model and its successes in the nuclear power industry, and expressed the hope that transnational industries such as oil might profit from INPO's model. INPO, which had long operated under the public's radar, now chose a different tack. INPO's highly praised performance on a public stage gave the INPO model greater credentials among WANO members.[39]

INPO's response to the Fukushima accident was, in part, the product of Ellis' frustrations with WANO's inability to hold its members as accountable as INPO could. Felgate admired Ellis, a graduate of the US Naval Academy, decorated fighter pilot and veteran of the Vietnam War and Operation Desert Storm in response to the Iraqi invasion of Kuwait, who rose to the rank of Admiral and Commander of US and NATO forces in combat and humanitarian operations during the war in Kosovo in 1998–1999. When Ellis retired from military service in 2004, he commanded the US Strategic Air Command at Offutt Air Force Base in Omaha, Nebraska. INPO, which had been struggling with succession issues for several years, selected Ellis as its president in 2005. Ellis embraced the internationalisation of nuclear power. As a military officer, he had served in all parts of the globe. From his perspective, as the nuclear industry became more global, US and foreign nuclear power operators alike saw greater value in fostering closer international connections and actions through INPO in what Ellis called "a coalition of the committed and willing." He had not witnessed such a deep commitment among many WANO members. In his mind, it would have been disingenuous to operate under the WANO banner, implying that WANO had capabilities it lacked entirely and adding "confusion to an already complex environment. If we were to label ourselves 'WANO'," Ellis noted, "our task would have added confusion, organisational complexity and inefficiency to an environment that already had far too much of all three."[40]

In addition, INPO had revved up its Partnership for International Nuclear Safety (PINS) programme, an outgrowth of the International Participant Advisory Committee created by INPO in the early 1980s. Stricker and Ellis had discussed the activity privately just after Fukushima, and the two men agreed that its activities should be made clear to the WANO Governing Board. Ellis explained that PINS had grown from the recognition by some international participants of the need for a higher level of integration in the nuclear community for services that WANO did not provide. The mission of PINS was to "set the global standard for nuclear safety" by demanding excellence of the members and expecting it of others. He stressed that the role of PINS was "complementary" to WANO and that a criterion for inclusion was full participation in the association's programmes with the objective to strengthen WANO. Key expectations of PINS members were a systematic approach to training, which WANO did not offer, and peer reviews every two to three years.[41]

Stricker was aware of the WANO-INPO tensions. At his invitation, Ellis began attending WANO Governing Board meetings at the end of 2009. The move proved to be a double-edged sword. It gave the head of INPO direct access to the Governing Board, but it also made him witness to its foibles. His tough questioning, Felgate observed, "pushed WANO forward more than any other board member". But at the same time, Ellis grew increasingly frustrated to "see things not getting done that should have been done in a more timely manner, decisions made by WANO for which there was no implementation or follow-through among other members, and the lack of participation in Board meetings by CEOs from other countries". All the self-assessments of WANO had said much the same thing – "there's a lack of commitment by our members and there's a lack of follow-up and follow-through by

WANO when problems are identified." Not every member was fulfilling its member obligations, and Cavanaugh had made the theme a central part of the Chicago BGM in 2007. There had been much talk, yet in Ellis's view, little had changed. Ellis also demanded accountability, but he held an additional card. If WANO couldn't respond, INPO could – and would.[42]

WANO did not have the resources to mount a response anything like that of INPO, but the WANO Governing Board immediately met via a conference call to offer what assistance it could. The association also received daily updates from Tokyo Centre that were passed on to member CEOs. At WANO's request, Professor Vladimir Asmolov, first deputy director-general of Rosenergoatom, travelled to Tokyo to head a WANO team to determine members' capabilities for support and assistance. Asmolov had extensive experience at Chernobyl following the 1986 accident, an ideal background for leading the team, but he was not well received in Japan. WANO also revised and reissued a SOER prepared by INPO asking members to evaluate and respond to specific issues related to their readiness to mitigate a beyond-design accident. In an unusual move, Stricker also recommended that members share the SOER with their regulators, part of an effort to urge the International Atomic Energy Agency to cooperate with WANO. Finally, Stricker proposed the formation of a high-level Post-Fukushima Commission to recommend changes to WANO.[43]

The Governing Board listened carefully to a report on the reaction of Atlanta Centre to WANO's response at a meeting in Paris in March 2011. Duncan Hawthorne, a WANO governor from Atlanta Centre, explained that at a meeting the previous week, the centre's governors had stressed that the WANO response was largely ad hoc, something they found unacceptable. The Fukushima calamity exposed the fact that WANO did not have an emergency plan to define its actions and responsibilities in such an event. Hawthorne stated that while it had been appropriate to send member

CEOs information on the accident, "many did not appear to have distributed it within their organisations". As a result, he said, WANO was "unjustly accused of not communicating" and, in light of the failure, he recommended a re-evaluation of the distribution list. But another issue also bothered the Atlanta Centre Governing Board: the paucity of knowledge about the Fukushima plant, which had not had a peer review since 2003, several years beyond WANO's announced goal of every six years. The frequency of peer reviews, Hawthorne argued on behalf of his Atlanta colleagues, needed to be increased.[44]

Once the floor opened for discussion, so did the floodgates of frustration with both WANO and TEPCO. Given what had occurred at Fukushima, peer reviews should include design review, emergency preparedness and accident mitigation in the future, several governors suggested. Others proposed a major overhaul or reorganisation of WANO and its Charter with an emphasis on safety and member commitment. For Anatoly Kirichenko, the Director of Rosenergoatom's Department for International Cooperation and Foreign Trade Affairs, the department responsible for the sale and export of Russian-designed and -built nuclear plants, Fukushima and the industry's response were crucial for future business. He admonished TEPCO for its lack of information and transparency. Moreover, there was another problem – WANO was "too bureaucratic and must be significantly improved or recreated". Like some others, Kirichenko called for revising the WANO mission statement. The governors suggested that WANO become more visible and craft a response to the accident. "WANO does not exist in the eyes of the public...because we have said nothing [about Fukushima]," said one. "We must speak." Transparency was essential to prevent a public lack of confidence.[45]

All the governors agreed that a strong WANO was essential to the industry's future. Fukushima could become the catalyst for engaging CEOs to recognise their responsibilities. "If every CEO is not substantially engaged, we will have failed," predicted one governor. Tom Mitchell, CEO of Ontario Power Generation, told the governors that "we are at a crossroads and WANO needs to be recreated or restructured to a position of strength." Ellis, for all the frustration with WANO, nonetheless firmly believed in its benefits and was determined to see it advance. He stated that Fukushima and its aftermath was "a transformational time for WANO when it is important to think clearly what WANO is about and why it was created. What we do will shape the future of WANO for better or worse."[46]

Calling for significant additional revisions to the WANO Charter, even before the major changes adopted in New Delhi had sufficient time to be implemented, indicated how badly Fukushima had shaken the confidence of WANO's governors. The organisation's response to Fukushima, one noted, "is WANO's second and last chance to get it right".[47]

As a result of the suggestions offered at the Governing Board meeting, Stricker created a Post-Fukushima Commission to examine the accident and WANO's response and to make recommendations to improve the association's programmes and strengthen its organisational structure. He appointed Mitchell chair of the Commission. An American with an undergraduate degree in nuclear engineering from Cornell University and a master's degree in mechanical engineering from The George Washington University, Mitchell had run nuclear plants in the US and Canada and held several top positions at INPO, including vice president of the international division. He had made a name for himself among the nuclear community by taking over operations of the Peach Bottom Nuclear Power Station in Pennsylvania when it was shut down by regulators and had established the plant as a recognised leader in safe and reliable operation. Mitchell's vigorous support for strengthening WANO made him an ideal choice.[48]

Other key members represented the other three centres, including Vladimir Asmolov, Deputy Director-General of Rosenergoatom, from Moscow; Bill Coley, former President of Duke Power and former CEO of British Energy Group plc, from Paris; and Takao Fujie, President and CEO of JANTI, from Tokyo. In all, the Commission consisted of 14 CEOs and senior utility executives representing 12 countries. Its charge was not to investigate the accident but "to focus on the broad context of the accident and the implications to WANO". Stricker and the Governing Board believed that, to be most effective, the report needed to be completed before WANO's 11th BGM in Shenzhen in October 2011.[49]

Over the summer the commissioners held five meetings – in Atlanta, Paris, Seoul, Prague and Tokyo – travelling to every region to gather members' perspectives and to discuss the event with Japanese utility staff. Reporting on the Commission's preliminary findings in July, Mitchell told the Governing Board that "the post-Fukushima world is very different from before the accident. We can now consider that the new world requires a new WANO with a new level of capability and consistency." Nevertheless, Mitchell was a realist. He thought the governors would find some of the recommendations "aspirational, difficult, long-term and provocative". He warned that all the commissioners did not agree with all the proposals, but they did recognise that achieving full implementation would "take a large number of resources, extensive executive effort and well-thought-out implementation plans to accomplish".[50]

In the wake of Fukushima, the Commission judged that WANO's scope was insufficiently broad and the quality of its products and services sub-par. The Commission also worried that the organisation's credibility was not universally acknowledged by its members and that it lacked the degree of visibility to be credible in public situations. The report was, in the words of an industry commentator, "a sober but necessary assessment."[51]

While the Commission concluded that the Fukushima accident "was not a failure of WANO", it did "point out some gaps in existing WANO activities". The Commission identified a number of focus areas for WANO's consideration. Several were in direct response to Fukushima, but others were geared to strengthen and restructure WANO. One was to expand the association's fundamental premise to not only prevent core damage but also to mitigate and respond to beyond-design basis events, including off-site radiation releases. A second suggestion was to expand WANO programmes to include design implementation of safety fundamentals to prevent fuel damage and mitigate off-site radiation and public impact. The Commission also recommended a worldwide integrated nuclear event response strategy with "clearly defined roles, responsibilities and interfaces for WANO and other relevant organisations" such as the IAEA and the World Nuclear Association.[52]

Fundamentally, the commissioners believed that to carry out the association's responsibilities, WANO "must be credible, visible and internally consistent and effective". To accomplish this, the Commission suggested ranking plants after a peer review, with the bottom ranking being "unacceptable" – a practice that had proved successful at INPO. Finally, Mitchell told the Governing Board that WANO should conduct internal peer reviews on each regional centre to identify both gaps and best practices. Changes outlined in the Commission's recommendations would "require a level of internal consistency in implementation of expectations across all regional centres", he explained. "Gaps have existed for many years between regional centres. Closing these gaps must be accelerated." It was essential, Mitchell concluded, for WANO to develop "a mechanism to provide strong motivation for members to improve".[53]

The governors engaged in a wide-ranging discussion of Mitchell's presentation, questioning the reasoning for some of the recommendations, particularly plant rankings. If implemented, one governor mused, the rankings might tear Tokyo Centre apart. Another expressed great discomfort with a toughening of accountability in response to member performance shortfalls or "inadequate responsiveness". Stricker disagreed. "There must ultimately be consequences for a member that does not meet its obligations," he stated. WANO must have "teeth" to enforce its programmes. Others advised a slow, cautious approach to change, perhaps splitting the Fukushima focus areas from the other areas. But whatever the recommendations of the Commission's final report, nearly all agreed with their colleague Dominique Minière, Executive Vice President of EDF and head of its nuclear operation division, that "there must be a new WANO or there will be no WANO at all."[54]

By September 2011, a draft of the report was sent to the WANO Governing Board, making recommendations in five areas. As a result of Fukushima, the report recommended that WANO "must expand its present focus on the prevention of events to also include mitigation should events occur". Other recommendations identified needed improvements to address long-standing performance gaps within WANO. The Commission's key conclusion was that "WANO must change and it must change in some detailed and fundamental ways." A reasonable estimate to implement the changes recommended by the Commission, Felgate estimated, would be a tripling of resources.[55]

Not surprisingly, Fukushima dominated the time between the accident and the Shenzhen BGM, "the most significant seven months in the history of WANO," Stricker said. The London staff put in full days and then some, assisting the post-Fukushima commissioners while scrambling to reshape the Shenzhen agenda. The initial theme for the meeting had focused on the implementation of the New Delhi

BGM initiatives, the obligations of membership, members not meeting peer review schedules and, subsequently, member accountability. The events at Fukushima overwhelmed the planned agenda. With little time to prepare "all the spit and polish of a normal BGM," Shenzhen would become a working meeting – a "close the doors and tell the press we're sorry but this is a special BGM," Felgate recalled. With the Post-Fukushima Report as a roadmap, members would "roll up their sleeves and work out what WANO needs to do to change because of Fukushima".[56]

The Commission's charter had been broad: it could recommend any changes "it determined important to close existing WANO performance gaps, including changes to WANO programmes, processes, membership, governance or structure". The report admitted that the industry and WANO, in particular, were ill-prepared to support TEPCO because "WANO's full focus since its formation has been accident prevention and no procedures were in place to address nuclear response or mitigation." The commissioners recommended that the Governing Board expand WANO's scope to include emergency preparedness and severe accident management, including procedures, training, and readiness. WANO should not enter the design review business, but members should perform periodic assessments of plant design risks to "consider new information, operating experience and site characteristics", but otherwise offered no guidelines for implementation. Finally, WANO should take an active role in "promoting and implementing" a worldwide integrated nuclear industry event response strategy.[57]

The implications of Fukushima for the industry and WANO formed the second part of the Commission's recommendations. The credibility of the industry had been damaged, the report stated, and the political reaction in some countries was "intense and strongly negative". To its credit, the commissioners wrote, at the BGM in New Delhi WANO had initiated a "number of significant changes to its governance, mission, process and expectations. The Commission's work validates the wisdom and soundness of these changes." Once fully implemented, these revisions would "form a strong foundation for further improving WANO's usefulness and effectiveness".[58]

In addition to the changes adopted at the New Delhi BGM, the Commission suggested other structural improvements for adoption in Shenzhen, urging members to reinforce their commitment to WANO and provide the resources to complete the changes started in New Delhi and recommended in the Commission's report. The keystone to the Commission's recommendations was toughening the scope and frequency of peer reviews – one every four, rather than six, years with a follow-up review two years later – combined, for the first time, with performance rankings in relation to other plants. The Commission recommended that every WANO member utility receive a corporate peer review every six years. Based on the results of those reviews, the Governing Board could establish "an appropriate frequency" for future corporate peer reviews. In addition, start-up peer reviews conducted at each new nuclear power plant before it reached initial criticality were added to the WANO mission. Each WANO peer review should "clearly state" if the plant did or did not make sufficient progress in responding to areas for improvement (AFIs) from previous reports. In addition, each nuclear plant undergoing a peer review "should have an assessment assigned that captures the overall safety risk represented by the peer review report", the Commission urged, a first step toward adopting plant rankings. The commissioners wanted the Governing Board to be alerted to every nuclear power station whose performance was unacceptable, or where significant AFIs remained uncorrected, significant events were unreported or its WANO commitments were unmet. "A firm policy should be established in such cases for escalated actions, [including] whatever means is necessary to achieve an appropriate member response."[59]

The Post-Fukushima Commission also resurrected a longstanding frustration – inconsistency in the implementation of WANO programmes across the four regions as well as performance gaps in the regional centres and London. The Commission proposed conducting periodic peer reviews, nominally every four years, of each centre and the London Office "to measure the quality, effectiveness, efficiency and consistency of implementation of WANO programmes and results achieved." The results of those reviews, the Commissioners stated, should be reported to the Governing Board and regional governors and summarised for the members at the BGM.[60]

The Commission insisted that WANO should create an internal-to-WANO member system to report "important events (particularly of media interest) within hours" that included information on the affected plant with a focus on "known facts, informed initial judgements and usability to defuse misinformation." The report urged WANO to become more visible and to quickly implement the recommendations adopted in India and those of the Post-Fukushima Commission. The recommendations could be put into effect "without any additional changes to the current WANO mission, governance or structure beyond what was done" at New Delhi. Although adoption of the Commission's recommendations "will improve public confidence and respect for WANO", this could only occur with the commitment of each WANO member "to fully participate and support a more effective WANO", the report concluded. The Commission's message was clear: "WANO must change and it must change in some detailed and fundamental ways."[61]

But, as throughout WANO's history, the necessary changes could not occur without adequate staffing and resources. Shortfalls in staffing had long been affecting WANO's ability to meet member expectations and improve plant performance. The governance changes adopted in early 2010 had laid the groundwork for addressing the resource shortfall. Fukushima highlighted the urgency and immediacy of that plan. The goal was to increase the core staff at each centre, increase continuity of programme staff and reduce the number of seconded staff. In order to meet the Commission's recommendations, the Executive Leadership Team, consisting of the managing director and the four regional centre directors, developed a best estimate of the staffing resources required. The team calculated that it would be necessary to nearly triple the total number of WANO centre staff and secondees from 131 to 384 by the end of 2014. The team anticipated the largest growth at the Paris and Moscow Centres, which would more than triple in size, followed by a doubling of Tokyo Centre staff, though many considered Tokyo Centre's numbers to be inadequate. The London Office and Atlanta Centre would also add staff, but at lesser rates. While the numbers were estimates, their adoption at the Shenzhen BGM would represent a major new commitment of resources and a decision by WANO members to make the best of a second chance to keep their organisation viable.[62]

The WANO Governing Board approved the Post-Fukushima Commission's recommendations on 23 October, as the 11th BGM got underway. The next afternoon at an Extraordinary General Meeting the changes were presented to WANO members for ratification. Some 600 participants from 34 countries representing 152 different companies gathered at the InterContinental Hotel in Shenzhen, a city located just north of Hong Kong on the Pearl River in Guangdong Province. The delegates, "markedly more sombre than [in] previous years" given the gravity of the events at Fukushima, unanimously endorsed the Governing Board's action. "The members of WANO," Stricker announced, "have cast their vote and pledged to increase their commitment to nuclear safety in the face of the biggest challenges the nuclear industry has confronted in 25 years." Officials believed the rapid adoption demonstrated the value of the governance changes completed in 2010 – expanding the WANO Governing Board, improving direct CEO involvement and realigning primary membership from countries to operating companies. The vote also marked a fundamental shift in

WANO from "simply accident prevention to prevention and mitigation".[63]

In announcing the changes to the media, Stricker explained that while the Fukushima accident "was not a failure of WANO" it had exposed "some gaps in existing WANO activities, such as emergency preparedness, severe accident management, on-site fuel storage, and, to some extent, design issues." Now it was up to WANO and its members "to deliver on the commitments". There was hard work ahead, Stricker admitted, but he hoped to be able to complete the changes by the 2015 BGM. Implementing the recommendations, Stricker stated, "will result in a stronger, more effective WANO and nuclear industry".[64]

WANO would have a tough road to travel. Admitting a degree of failure in responding to Fukushima was an important first step. The recommendations unanimously approved at Shenzhen, intended to increase WANO's relevance and value, were the second stride forward. But "success", one observer commented, "will hinge on whether the 'new' WANO can effectively carry out a more ambitious mission across a diverse membership. In other words, will the tiger have enough teeth?" While a unity of purpose existed among WANO members, the organisation's history had demonstrated that achieving a true consensus across the four regional centres – each representing "different safety, performance and business constructs" – would be extremely complex and challenging. Nevertheless, if WANO and the nuclear power industry were to advance beyond Fukushima, all of the recommendations had to be embraced and enacted. Every nuclear utility had to unite in a concerted effort to eliminate any weak link in the chain of power plants and constantly improve operations, ever mindful of the danger of Carlier's volcano.[65]

LAST CHANCE **TO GET IT RIGHT**

Chapter Seven

ONE **WANO**

In 2012, one year after Fukushima, nearly 50 countries were operating, building or considering nuclear power facilities. In addition, more than 60 nuclear power plants were under construction in China, India, Russia, South Korea, France, Finland and the United Arab Emirates. China's nuclear plans were the most ambitious, and on track to triple the country's nuclear capacity from 12 gigawatts to 40. Such rapid expansion demanded enhanced nuclear safety on an international scale, said Pierre Gadonneix, Chairman of the World Energy Council, echoing a recommendation made soon after the Fukushima disaster by French President Nicolas Sarkozy. Gadonneix believed this goal could be achieved through the cooperation and coordination of the IAEA, WANO, the International Nuclear Regulators' Association and the International Nuclear Safety Group, among others. "The safety of global nuclear power," he wrote, "is one of the rare issues on which an international accord could be achieved… The need to act is urgent, and the time is right."[1]

WANO did not disagree with Gadonneix's goal. Nuclear safety had no borders. Since its founding, WANO had worked closely with the IAEA, coordinating its peer reviews so as not to interfere with the IAEA's own reviews. If anything, Fukushima had demonstrated the importance of collaboration. By autumn 2012, WANO and the IAEA had signed a new memorandum of understanding (MOU) that enabled the organisations to work more closely to support nuclear plant safety and reliability and to enhance the exchange of information "on operating experience and other relevant areas". WANO would communicate with the other regulatory bodies, but it would not share confidential member performance information with them. The MOU was similar to previous agreements, such as coordinating OSART and WANO plant reviews, but added exchanging information on a "serious event" at a nuclear power plant or fuel cycle facility. The agreement also provided for the exchange of staff on IAEA and WANO review teams and an exchange of documents on operating experience. WANO Chairman Laurent Stricker viewed the new MOU as an "important lesson we learned from Fukushima, the need for WANO to be better connected to and engaged with the IAEA".[2]

To meet the changing needs of the nuclear power industry, WANO's work went beyond the IAEA memorandum. Throughout its history WANO had been an organisation of experienced operators. WANO would continue to offer the technical support missions that were so important to increasing the competencies of these members. But WANO was embarking on a new era of nuclear history consisting of many new entrants on one end of the spectrum and, as the nuclear fleet aged, an increased number of decommissioned plants. The Governing Board recognised that WANO had to continually adapt and grow to meet those changes. In 2011 WANO opened a satellite office in Hong Kong to support pre-start-up reviews for new entrants in Asia.[3]

****

Although WANO had just completed an internal peer review of each of the regions and the London Office, the Post-Fukushima Commission urged another review to analyse internal WANO programmes for competency and consistency across the regions. The Commission concluded that the "nuclear industry had changed unalterably" and WANO "must be much stronger and have 'teeth'" to operate effectively in the new environment. To bring all the WANO programmes into closer alignment, to achieve consistency across the regions and to realise greater member commitment and accountability could help accomplish this goal.[4]

Stricker appointed Matt Sykes, a Chief Nuclear Officer of EDF Energy, which had become part of the French utility EDF when that company bought British Energy in 2009, to head the assessment team. The team consisted of four regional coordinators and 13 reviewers representing all the centres, supported by staff from the London Office. Between spring and autumn 2012, the assessment teams spent countless hours conducting the five reviews, identifying "considerable inconsistency in the implementation of WANO programmes across the four regions, as well as performance gaps in all the regions and London." Most of the activities in the Atlanta Centre "are performed well", the reviewers acknowledged, and only a "few items were identified [which could] further strengthen programmes". Paris Centre, too, received accolades for increasing its staff and starting several new initiatives. However, the reviewers cautioned that member performance gaps should be closed and that "the Centre needed to be more intrusive to move members toward excellence." At the Moscow Centre, the team "found considerable positive momentum to change and improve", though several programmes required more attention. The team also noted that "there is a lack of transparency by some members of Moscow Centre" that needed correction.[5]

Tokyo Centre, however, did not score very well. There, the team found "significant weaknesses in many aspects of the Centre's activities", the primary contributor being the "lack of support and engagement by WANO members affiliated with the Tokyo Centre". Tokyo's gaps fell into three categories – deployment of WANO programmes, strategic planning and leadership direction, and engagement and support by members. Many of the Tokyo Centre's weaknesses were "of long standing and…there is a lack of detailed planning". Of particular concern was the Centre's inability to "identify and address member performance issues".[6]

The findings of the assessment team regarding Tokyo Centre echoed the remarks of Zack Pate in recalling WANO's history in a 2012 letter to his friend Thomas N Mitchell, the President and CEO of Ontario Power Generation (OPG) who had chaired the Post-Fukushima Commission. The Tokyo region "has been the weakest of the four regions pretty much throughout WANO's history", Pate recounted. "The lower seniority of the Tokyo Centre Board members is one of the reasons."[7]

The London Office also received a dose of criticism. For the past three years, since the change of governance and membership approved at the 2010 New Delhi BGM and later as a result of the Fukushima accident, London had focused on Governing Board support, nuclear utility CEOs and stronger ties with other nuclear support organisations in order to gain member support and improve CEO engagement. However, "this focus upward and outward has resulted in too little focus internally," the report stated. For example, London had assumed little direction or "oversight" for WANO programmes, and there "was a lack of coordination among the four programmes." Management systems had to improve if the London Office were to triple in size, as "the informal processes that were suitable for a staff of 12 are not adequate for a staff of 36."[8]

In faulting London for the uneven levels of implementation of WANO's core programmes across the centres, the self-assessment team returned to the need for closer integration of the regions. Although guidance for WANO's programmes had been outlined by London and approved by the Executive Leadership Team, regional implementation was inconsistent. In some cases this was because the "guidance is not clear, is open to interpretation or simply does not exist. In other cases, the freedom given to regions to implement WANO programmes based on specific cultures in their regions, has led to deviations from approved guidance." The primary cause for this failure was "insufficient oversight and monitoring of programme effectiveness by London". In the office's defence, however, the team concluded that earlier attempts initiated by London and the ELT to close these gaps had ground to an abrupt halt when nearly all of WANO's attention was directed toward responding to the Fukushima event. Nonetheless, greater oversight was an important target for the London Office to achieve if WANO and its core programmes were to function efficiently.[9]

WANO's founders had stressed that the industry was only as strong as its weakest performer. "We are hostages of each other," Bill Lee had told members at the May 1989 WANO Inaugural Meeting in Moscow. Yet, apart from members of Atlanta Centre, over the years the other regional centres did little to identify and bring corrective action against the stations that posed the greatest risk to nuclear safety. The causes were varied, according to the report. Tokyo Centre believed that its members were "self-reliant and responsible for their own performance, and there has been a reluctance to be perceived as too intrusive or assertive". In fact, the reviewers pushed for more intrusion into the operations of member plants for both Tokyo and Paris Centres. In Moscow Centre, the reviewers noted that "the quality of peer reviews at some plants [was] reduced by their lack of openness, limited access to staff and physical spaces, and excessive and unwarranted challenging of issues." Even without a formal ranking system, everyone knew which plants had performance gaps and posed a safety risk, the reviewers stated. It was WANO's responsibility to focus discussion on plant performance and insist on improvement, casting aside worries that a list of "focus plants" would, in the wake of Fukushima, cause embarrassment or adverse public reaction. A formal ranking and response procedure, not unlike INPO's, would greatly strengthen WANO's drive for excellence.[10]

The major complaint of Atlanta Centre was that nuclear plants in the US "make almost exclusive use of INPO (rather than WANO) for technical support and professional and technical development". Nevertheless, the WANO Internal Assessment Report was a plea for greater standardisation of recruitment and training of staff based on INPO's methods. The recommendation for plant rankings was also drawn from the INPO model. After more than two decades, WANO, which had integrated some INPO programmes into an international framework, found that it could best survive by adopting many other INPO features. Peer reviews, initially rejected at WANO's creation, were now critically important to its success. But peer reviews, without peer pressure applied evenly across the organisation, lacked clout and value, as INPO had demonstrated. Full top-level engagement, higher-quality programmes, sufficient resources, operational accountability and continuous improvement in all areas were required to ensure WANO's long-term viability and success. The challenge for the Governing Board would be finding the "right balance" between local interpretations of WANO policy and the high degree of integration, standardisation and centralisation needed to achieve the goals of the New Delhi and Shenzhen BGMs. Only a "One WANO"—not four confederated regions—could remain effective in a shifting nuclear world.[11]

****

Whatever the "right balance" might be, three items demanded WANO's immediate attention. The first, an expansion of WANO's responsibilities after the 2010 New Delhi BGM, was the creation of a pre-start-up branch office of the London Office in China to assist the large number of Asian nuclear plants scheduled for such reviews later in 2012 and 2013. Not all regional governors favoured this initiative, unwilling to cede more control to London. Others viewed pre-start-up reviews as crucial to WANO's future since a couple of events had occurred at units that were relatively new. When initial attempts to locate the office in Beijing failed, WANO negotiated locating it in Hong Kong. WANO Managing Director George Felgate later characterised the new office as "one of my most difficult accomplishments". By September the Hong Kong branch of the London Office had opened, but would not be fully staffed for another year. The London Office, too, was in flux, relocating from a "shabby location in west London", as one WANO official described it, to more spacious and open quarters in a modern high-rise office building in the redeveloped Canary Wharf area of the city. The new space tripled the London Office's square footage, allowing the planned expansion in staff to occur while keeping rental costs in check. The move was "symbolic of a new WANO", Felgate believed, of an organisation "looking to the future".[12]

The second was a crisis in the safety culture of Korea Hydro & Nuclear Power Co, Ltd. (KHNP), the South Korean nuclear operator. The event occurred in early February 2012, during a planned outage at the company's Kori 1 unit, South Korea's first nuclear reactor that had been operating since 1978. During the outage, the plant lost its outside source of electrical power and a back-up diesel generator failed to start. The incident was not reported to authorities for more than a month and WANO for months. However, WANO conducted an on-site review and issued a Significant Event Report. The IAEA also sent a mission to Kori. Both groups concluded that there were concerns with the company's safety culture. The company had failed to report incidents, failed to communicate with WANO and allegedly used counterfeit parts in some units. Stricker followed the WANO review with a letter to KHNP requesting the company to allow WANO to conduct a corporate peer review, but the company had not responded. The Governing Board was significantly concerned. Several WANO governors considered the situation "unacceptable" and argued that WANO's credibility was at stake. "It is time for WANO to step up," one governor declared. Others favoured an escalation of WANO actions. "If one member is not playing the game," one governor said, "WANO must react quickly with significant sanctions."[13]

Stricker was unwilling to push KHNP to the brink, much less over it. With the resignation of the CEO, the company restructured its corporate leadership, so it made sense for Stricker to ease the pressure and delay the review. Instead, he proposed sending another letter, requiring that the company schedule a corporate peer review at its "very earliest opportunity." Dr SK Jain, Chairman of the Tokyo Centre Board of Governors, also urged a cautious approach. He recognised that there were "limitations" on the centre's ability to work with KHNP, but he would go to Korea to meet personally with the head of KHNP to convey the WANO Governing Board's concerns and urge forceful backing for the corporate peer review and support mission. Jain said he wanted to "assure the governors that the region is putting pressure on the member". By the beginning of 2013, KHNP had agreed to host a corporate peer review by the end of the year.[14]

Jain's initiative occurred because he recognised that the Tokyo Centre was under some pressure. In his mind, the WANO reforms, with their emphasis on consistency and accountability, carried the seed for the potential disintegration of Tokyo Centre and its cultural adjustments to the implementation of WANO's programmes. There was a feeling of "distrust", Jain told his fellow governors, that "WANO is diluting the autonomy of the regional centres". And Tokyo Centre was "very sensitive" to this. Nevertheless, after Fukushima, the Japanese government closed all 52 of the country's nuclear units. Public trust in nuclear power was shattered in Japan and shaken in the rest of the world. Moreover, the KHNP issues further weakened the centre's stature. In the second half of 2012 the influence of Tokyo Centre in Japan was further diminished. Japanese utilities, reacting to Fukushima, responded by reshaping the Japan Nuclear Technology Institute into the Japan Nuclear Safety Institute (JANSI) and adopting many of INPO's programmes with an emphasis on accountability and peer pressure. The new JANSI would "serve as a powerful industry driver" with the added "autonomy of making judgments unaffected by the intentions of nuclear operators". Its mission was to "pursue the world's highest level of safety", including ranking of power plants and use of peer pressure among CEOs to achieve those goals. The organisation would also take on additional functions, including severe accident response and emergency preparedness. JANSI's broadened mission included specific changes that WANO was urging Tokyo Centre to accept. Stricker served as an adviser to JANSI, but Tokyo Centre played little or no role in the creation of the new Japanese organisation. However, at Stricker's urging, when Tokyo Centre sought to expand its offices, the Tokyo Centre governors considered moving into the JANSI office to bring the two organisations together, much like the INPO/WANO Atlanta Centre model. Equivalency to that model, however, was the key challenge if the new structure were to be successful.[15]

With Tokyo Centre reacting to both Fukushima and its internal shortcomings, the centre of nuclear power in Asia had shifted from Japan to India and China and their vigorous construction programmes. Some governors thought of moving WANO's regional centre out of Japan, wondering if there were other ways to boost WANO's effectiveness in Asia.[16]

To remain viable as an organisation, WANO's third focus and top priority, the governors concluded, must be the rehabilitation and reformation of Tokyo Centre. The WANO Governing Board pushed for CEO engagement and "the acceptance of peer reviews without defensiveness". All the other regional centres, the governors stated, "should assist during this time of difficulty," but it was up to the Tokyo Centre Regional Governing Board to make "permanent improvements". WANO urged the regional board to draft sets of intermediate and long-term plans to become a fully self-sufficient centre. In the interim, the WANO Governing Board agreed to augment the centre's staff, while the Managing Director and ELT drew up a strategic plan to provide WANO assistance to Tokyo Centre. The first step in this process was a presentation to the Tokyo Centre Governing Board by Matt Sykes of the internal assessment team, which had just completed an assessment of the centre. Sykes outlined specific areas to correct and improve. Jain reported that the assessment contributed "to the desire of the governors" to make those improvements.[17]

The obstacles faced by Tokyo Centre – lack of CEO involvement, language barriers and diplomatic problems, among others – were not new, as Jain explained to the WANO Governing Board. Yet, he noted, "action plans [had] been developed and endorsed by the Regional Governing Board." However, he cautioned that the regional governors were not CEOs or chief nuclear officers (CNOs) and that "even appointed governors send proxies." He thought it imperative that Tokyo Centre had to involve "the highest levels" for WANO to have a significant impact on the region. A Tokyo Centre governor asked the WANO Governing Board to give the region more time and to assist in providing restart reviews for Japanese plants as they came back on line.[18]

Jain sought to lead Tokyo Centre's recovery and, at the same time, resolve some of the thorny issues that plagued WANO's Asia region. A short, stocky man and articulate advocate for nuclear safety, Jain had been the Chairman and Managing Director of Nuclear Power Corporation of India Ltd and Bharatiya Nabhikiya Vidyut Nigam Ltd (BHAVINI), the group in charge of India's fast breeder reactor programme. Widely respected for scientific achievements and his bureaucratic shrewdness, Jain was elected to chair the Tokyo Centre Board of Governors in 2012 and immediately became the centre's chief spokesman and defender at WANO Governing Board meetings. He cared deeply about nuclear power's future in Asia, recognising that there was "a lack of commitment within the region and that some members do not appreciate the WANO programmes". Privately, he worried that changes in WANO's peer review system that included plant rankings would tear the Centre apart, perhaps causing the Japanese to drop out altogether. Yet he was determined to make WANO vital to the region. Although there appeared to be far too much on Tokyo Centre's agenda – reviews for long-term shutdown, ongoing issues with KHNP and pre-start-up and re-start-up reviews – Jain worked on plans to ensure that Tokyo Centre staff had the required competencies to deal with future needs. In addition, along with other WANO governors, Jain concluded that CEO involvement had to be strengthened, "especially with the forthcoming evolution of the Asian countries becoming the dominant continent in the respect of nuclear power". Indeed, if Tokyo Centre could not meet the needs of the region, it was highly likely that Asian members would join a centre that best represented their nuclear technologies or other needs, rather than geographic proximity. To prevent that from happening, Jain asked that each Centre detail two experts to Tokyo Centre to "demonstrate solidarity".[19]

As Jain and the WANO Governing Board sought solutions for reinvigorating the organisation generally and Tokyo Centre particularly, Stricker, the chairman leading these changes, was ending his term and stepping down. Prior to the election of a successor, the Governing Board extended the terms of the chairman from two to four years, and the managing director from two to three years, with the idea of making the managing director a voting member of the Governing Board. The board then elected Jacques Régaldo as the new WANO chairman to succeed Stricker on 1 March 2013. George Felgate, wishing to return to his family in Atlanta, had also announced his intention to leave at the end of 2012, and the search for a new managing director began. Once again, the succession continuity WANO sought eluded the organisation.[20]

****

The key to WANO succeeding in becoming "One WANO" was the time and effort put in by the chairman and managing director to personally convert wayward nuclear utility executives to their increased obligations. Pate warmly approved Stricker's approach. Mulling over WANO's history, he explained to a friend that "to be really successful, WANO must be a 'club of CEOs' and not a club of plant operators." Pate praised the emphasis on building relationships and ownership with CEOs and other top managers for each member.[21]

With his term about to end in little more than a year, Stricker reflected on what WANO had accomplished since the 2011 Shenzhen BGM. The self-assessment of each of the centres and the London Office had been completed and regional governing boards had begun to implement as many reforms as possible under the leadership of the Board Oversight Committee, headed by WANO President Professor Vladimir Asmolov. The regions were to report their progress at the Moscow BGM in 2013. Fukushima marked a turning point for the industry. Stricker was convinced that another catastrophic event threatened the future of commercial nuclear power and stressed that Fukushima had clearly demonstrated that "an accident in one country had consequences for all nuclear operators elsewhere." Accordingly, Stricker and Felgate aggressively preached the gospel of involvement, accountability and ample fiscal support to WANO members. The two men travelled extensively, visiting utility CEOs, urging them to embrace WANO's changes and renew their commitment to the organisation, including payment under the new fee structure and providing quality secondees. In his four years as WANO Chairman, Stricker's vision for WANO had shifted from an association based on regional autonomy toward a stronger, more unified, governing structure seeking consistency across its policies and programmes. Amid the swirling currents of organisational change, preparing for and conducting two BGMs within two years, the impact of Fukushima and the recommendations of the Post-Fukushima Commission, Stricker and his staff had set a course that would steer WANO in a new direction.[22]

****

The decisions of WANO members at the New Delhi and Shenzhen BGMs, implemented by the tireless work of Stricker, Felgate, the Post-Fukushima Commission, Matt Sykes and the self-assessment teams, the Executive Leadership Team and the Board Oversight Committee, had established a clear roadmap and firm organisational structure for WANO and the industry's best "last chance" to remain viable. There was much to be done to accomplish the directives of the Post-Fukushima Commission and to close the programme gaps among the four regions by opening of the 2015 BGM. With Stricker and Felgate departing, it was imperative that the new leadership embrace the changes and the challenges.

Once again, the nominations process for chairman was not without controversy. Within a few weeks of the election only one strong candidate had emerged – Jacques Régaldo, like Stricker, a senior executive of EDF. The dilemma was that, historically, the WANO chairmanship had rotated among the regions. The Americans pushed to maintain the tradition, proposing two candidates from the Atlanta Centre, only to see both men withdraw their names. At the eleventh hour, Atlanta urged Pate's friend Tom Mitchell, the president and CEO of Ontario Power Generation and chairman of the Atlanta Centre Governing Board, to run. Mitchell was a logical choice and he had superb credentials for the position. The Post-Fukushima Commission, which Mitchell had chaired, had largely defined WANO's direction after 2011. The problem was that Mitchell also chaired the Nominations Committee that had put forward Régaldo's name. Two weeks before the Governing Board was to select Stricker's replacement, to the surprise of many, Mitchell announced his candidacy. The debate among the members of the Nominations Committee was whether the chairman's job would be a part-time or a full-time post, as it had been under Stricker. Many governors believed that the position, especially with its expanded duties under the new governance arrangements, should be full time. Those who backed Mitchell, who, if elected, would continue to have significant employment obligations in Canada, lobbied for a part-time chairman. The vote divided along geopolitical lines, with Mitchell's support largely from North America and Régaldo's from Europe, Russia and part of Asia. Régaldo was elected. He would assume the chairmanship on 1 March 2013.[23]

Stoutly built with a round face and infectious smile, Jacques Régaldo was urbane and sharply intelligent. Born in Bordeaux in southwestern France in 1956, he was the youngest of three sons of a professor of French literature. From his father he acquired a lifelong interest in art, literature, and the humanities, even as he pursued a career in science. Régaldo graduated from the prestigious École Nationale des Ponts et Chaussées, France's premier engineering university, founded in 1747. Upon graduation in 1980, he joined EDF, first working at a coal-fired plant, then switching to the company's nuclear division, which was expanding during that decade. In his career with EDF, Régaldo served as Site Vice-President at two nuclear power stations, as Managing Director of the Fossil and Hydroelectric Generation Department, and as Executive Senior Vice-President of EDF's Generation and Engineering Division. He also headed the Employment Division of the entire EDF group in charge of recruitment, training and career management. Régaldo had participated on EDF nuclear inspection teams and was familiar with the peer review process. He had just stepped down from his position as Operating Senior Vice-President for the Generation and Engineering Division of the EDF Group when he was tapped for the WANO job. The new Chairman also had considerable international experience as a member of the Board of the British Energy Nuclear Group as well as Constellation Energy Nuclear

Group in the US, both companies being subsidiaries of EDF. While he joked that his first language was "nuclear operations", Régaldo's English was excellent. [24]

Régaldo had done his homework. He had met many of the governors earlier in his career through his work at EDF and with INPO and Japanese companies. He told the Nominations Committee about his two priorities. The first was to implement the recommendations of the Post-Fukushima Commission, a priority very much in line with WANO goals. The second one, he said, was to meet most of the members, including the most important members, two to three times a year. But he also wanted to make contact with "more isolated members or countries", such as Iran, South Africa, Pakistan and Armenia. "My idea was very simple. All the WANO members should be a full part of the community, not isolated for any reason." Equally important as his nuclear and managerial background, Régaldo would be a full-time Chairman. Moreover, EDF would cover his salary and other expenses, a major consideration at a time when WANO staffing and budgets were expanding to meet the recommendations of the Post-Fukushima Commission.[25]

A month before Régaldo became Chairman, Felgate announced to the Governing Board that he and the Oversight Committee were confident that the implementation of the Shenzhen BGM decisions regarding the Post-Fukushima Commission recommendations, some 12 projects, would be well in hand by the Moscow BGM in May 2013 and completed by the 2015 BGM. It was an ambitious schedule but, most governors thought, not impossible. With Felgate leaving and the schedule to finish before October 2015 unchanged, the selection of a new WANO managing director became a high priority. Eleven candidates applied for the position. The Nominations Committee, newly named the Strategy, Governance and Nominating Committee to reflect its additional responsibilities, winnowed the list and recommended several possibilities. As part of the process, each candidate wrote a paper outlining his views of WANO and how it should move forward. Over the next two months, Régaldo reviewed the submissions and interviewed the candidates to determine who might best complement his vision for WANO.[26]

Régaldo and the Governing Board selected Kenneth "Ken" Ellis, a Canadian who had worked for Ontario Hydro since 1981, primarily at its Bruce nuclear generating station on eastern shore of Lake Huron. Tall and angular, with short-cropped hair and a military bearing that bespoke his early career, Ellis was born in Espanola, a small pulp and paper town on the Spanish River west of Sudbury in northern Ontario. He graduated from the Royal Military College of Canada in Kingston, earning a degree in mechanical engineering. He spent four years as an aerospace engineer in the Canadian Air Force, primarily in search and rescue operations off the country's west coast. Search and rescue, he recalled, was "one of those professions where some days it's sheer jubilation and other days just sheer depression. It all depends whether you're actually a rescue or just a recovery." After four years in the service, Ellis opted for an industry career and joined Ontario Hydro's nuclear fleet, steadily rising in operations, maintenance and engineering management. Bilingual, in 1994 he became its liaison engineer, a kind of nuclear attaché, to EDF's nuclear inspectorate in Paris. For the next two years he visited numerous nuclear plants in France as part of EDF's internal control group. The posting was "a real learning experience. I saw things that they could improve upon and I saw things they were doing that could improve our operations." He also completed a directors' course at the University of Toronto, specialising in corporate administration and governance. Over his 31 years at Bruce, Ellis served as the station Vice-President of Bruce B, Site Vice-President of Maintenance, Site Vice-President of Engineering and Chief Engineer, and Executive Vice-President and CNO. Ellis's background seemed ideal to Duncan Hawthorne, the head of Bruce Power, and he suggested that Ellis apply for the WANO position.[27]

Ellis was intrigued. He held WANO in very high regard. Bruce had been the site for one of WANO's first pilot peer reviews in the fall of 1992. The experience was an epiphany for Ellis. "We realised we were not nearly as good as we thought we were in terms of standards, operational safety and a nuclear safety culture. It was a huge wake-up call that we had developed in isolation." That experience "made me a firm believer in WANO, having seen first-hand its impact and its effectiveness in improving nuclear operating safety".[28]

Ellis worked hard on his application submission, writing an extensive paper on WANO governance structure and which leadership attributes and techniques could best fulfill the managing director position. Ellis believed that leadership based on "collaboration, facilitation, support and influence would work more effectively within a confederation governance structure". In February 2013, Ellis flew to Paris to interview with Stricker and Régaldo. Stricker wanted to ensure the continuity of the changes that WANO was making; Régaldo wanted to be certain that he and Ellis would be compatible. Both men were satisfied that Ellis was the person for the job. Seven weeks after he was hired as Managing Director, Ellis was on his way to the Moscow BGM.[29]

****

Twenty-four years before, in 1989, WANO had held its Inaugural Meeting in Moscow at the very same place it met for its 12th Biennial General Meeting in 2013. Both the venue and the organisation had greatly changed, of course, recognised only, perhaps, by the site location on the Moskva River or the organisation's name. The internal working parts of both were vastly different. The Sovincentr of 1989 had been transformed into the Crowne Plaza Moscow World Trade Centre. WANO was no longer a fledgling industry nuclear safety association. It had changed most dramatically in just a few years, since initiating governance and membership changes in 2010 in New Delhi and implementing the recommendations of the Post- Fukushima Commission after 2011. WANO's internal changes were considerable: to expand its scope into emergency preparedness, on-site fuel storage and some design considerations; to improve the quality and frequency of peer reviews, be they plant, corporate, or pre-start-up; to implement a worldwide nuclear event response strategy; to become more visible and transparent; to conduct more frequent self-assessments; and to ensure the consistency of WANO's core programmes. Amidst the evolution of the organisation, WANO could not reduce its focus on the fundamentals of safe operation of nuclear power plants.[30]

To assist in this effort, WANO members ratified several crucial changes at the 2013 Moscow BGM, approving alterations in the WANO Charter and Articles of Association. The Charter was amended to include the managing director as a voting member of the Governing Board. At the same time the Articles were altered to extend the term of the WANO chairman to four years, with service of no more than six years unless decided otherwise by the Governing Board. The managing director's term was also lengthened, to an initial three-year term with options for additional two-year terms "or as determined by the Governing Board". Duncan Hawthorne, the President and CEO of Bruce Power in Canada, was elected to succeed Asmolov as WANO president. Hawthorne was an excellent example of the transnational utility executive. Born and educated in Scotland, over his lengthy career in power generation he held senior positions in the United Kingdom, United States, and Canada. Hawthorne had been active in WANO activities and publicly criticised its members after Fukushima. "We need to accept that there has been a lack of progress in several areas", he said in his distinctive Scottish burr, "and this cannot continue. WANO must expand its scope and capability to give its members both what they want and what they need". As the incoming WANO president, Hawthorne replaced Asmolov as the chair of the Oversight Committee. He told the members at the Moscow BGM that he would continue Asmolov's initiatives and focus "on ensuring the full implementation of all Shenzhen actions by the 2015 BGM".[31]

The challenges of fulfilling the tasks set forth in the Post-Fukushima Commission report were daunting, but WANO, which had failed to implement such changes in the past, had acquired a new sense of urgency after 2011. In his outgoing presidential address, Asmolov concluded that WANO was undergoing "a transition from the stage of discussion to the stage of realisation". There were "no immediate answers to the questions we face," he added, but the organisation now had the opportunity to finalise "the job of turning a new page in WANO's life".[32]

****

To reach that new page, Régaldo and Ellis agreed at the outset that, unlike Stricker and Felgate, they would travel separately – "divide and conquer," as Ellis good-humouredly described it. Stricker and Felgate had travelled together during their campaign to court all the CEOs and get the regions more involved with WANO. Régaldo decided to continue those efforts on his own, believing that he and Ellis could cover more ground working separately. Moreover, Ellis concentrated on restructuring and building the staff of the London Office so it could provide tighter oversight and governance of the four core WANO programmes, as well as narrowing the gaps and inconsistencies existing among the regional centres as outlined by the Post-Fukushima Commission. To implement the governance reforms mandated at the New Delhi BGM but derailed by the Fukushima Daiichi disaster, Ellis met with the Executive Leadership Team and developed a set of standard policies to bring greater consistency to the programmes. In many respects, Ellis believed, the London Office "was a bulldog without teeth". He would try to build a consensus first, but if that were not possible, the London Office would set the course. Important innovations in this direction were developing a social media strategy and taking over governance of WANO's programmes. Ellis placed added importance on communications with members, making communications a fifth WANO core programme and promoting Claire Newell to Communications Programme Director. Together, they overhauled and redesigned the WANO website, adding interactive components, and turned *Inside WANO* into a digital publication delivered to desktop or mobile devices. "Given the importance of our mission", Ellis said, "the need to tell our story and ensure our programmes and products are well understood has never been greater". Communications would be "the new face of WANO".[33]

Other crucial changes were also occurring. Within a year, the WANO Governing Board took a crucial step toward enhancing the status of the London Office and the efficiency of the association when it changed the name of managing director to chief executive officer of WANO, a title that was more in keeping with the changes in governance and the powers of the London Office. In addition, as WANO placed greater emphasis on the active commitment of utility CEOs, the administrative head of the association would be an equal at the Governing Board table, both in title and as a voting member.[34]

Without a fully staffed office, London could not provide the oversight capabilities identified by Sykes and his team in the 2012 assessment report and now expected of it. Ellis informed the Governing Board that he would approach the regional centres for assistance in filling the vacancies. By the fall of 2013, Ellis announced that the hiring campaign had been successful and that London had doubled its staff. To expand the office's oversight of WANO's four core programmes, Ellis brought in full-time directors for each. Along with their governance duties, the general goal was to improve the core programmes by getting the centres to work together as a team to achieve consistency, as well as bolster oversight pressure from the London Office.[35]

A key challenge was to bring Tokyo Centre into alignment with the other centres. Jain, the chair of the Tokyo Centre Governing Board and a powerful advocate for strengthening WANO in Asia, told the governors that Tokyo Centre was not running efficiently. He reported that only one CEO served on the Tokyo Centre Governing Board and that there was "a lack of commitment within the region and that some members do not appreciate WANO programmes". This situation had to change. WANO leaders recognised that Asia was fast becoming the dominant continent in respect to nuclear power. The future structure of WANO had to incorporate this change in order to provide effective support and safety management to its members. In the spring of 2013 much of the discussion among the WANO governors revolved around strengthening the experience and capabilities of the centre's staff and the level of CEO involvement. Paris Centre proposed sending two people to Tokyo and suggested other centres do the same. "Budget should not be an issue when safety is at risk," the representative said. A Moscow Centre governor agreed, stating that WANO should look at the centres collectively, not individually. Sending skilled individuals would help ensure consistency among the WANO programmes.[36]

By the end of 2013, the situation in Tokyo Centre had improved. Importantly, a member of the WANO Board of Governors, Makoto Yagi, the CEO of Kansai Electric and Chairman of the Federation of Electric Power Companies of Japan (FEPC), resolved to tackle the task of CEO involvement. He explained to his colleagues that Japanese nuclear operators had been dealing with issues stemming from Fukushima rather than with the issues of Tokyo Centre. Ellis, Régaldo and Yagi initiated a small group meeting of CEOs to "help them understand the issues that WANO currently faces". Régaldo and Ellis explained WANO activities and the problems facing Tokyo Centre. The meeting was a turning point, and the Japanese CEOs agreed to serve as governors on the Tokyo Centre Governing Board. After that meeting, Yagi told the WANO governors that he could assure them that "in the future the [CEOs] will strongly support Tokyo Centre." By January 2014, all the Japanese CEOs had joined the regional board. In addition, the CEOs promised to provide additional resources for training peer reviewers and to send more experienced secondees to serve at Tokyo Centre. Yagi was also optimistic about the centre's role in upcoming peer reviews, stating that the centre would have five team leaders in 2014, an improvement that would narrow the gaps in the peer review process between Tokyo and the other regional centres. A further step would be completed in the late spring of 2014 when Tokyo Centre moved into new offices near JANSI, in a space that could eventually accommodate a staff of 100. With the support of the other centres, he said, "all efforts would be made to rebuild Tokyo Centre quickly."[37]

The Governing Board continued to assist Tokyo Centre, both financially and with seconded staff. By the end of 2014, Tokyo Centre had reached its targeted staffing levels for the year and boosted its funding of English language lessons and staff training for peer reviews, including a mentoring programme for less-experienced reviewers. In addition, with the assistance of secondees and the other regional centres, the centre had established a team leader/reviewer qualification plan to increase the quality of the region's peer review representatives. The centre also offered a plan to place a site representative at Fukushima Daiichi to ensure that WANO remained engaged with the station and to oversee the progress at the plants.[38]

In July 2015, WANO conducted a second follow-up assessment of the Tokyo Centre. The assessment noted progress in several areas, including continued improvement in member CEO commitment and involvement. The Tokyo Centre Governing Board approved a Site Representative (SR) programme, aimed at assigning an experienced WANO SR for each station to improve Tokyo Centre's ability to monitor plant performance and provide effective assistance. New leadership at Tokyo Centre implemented a new strategic plan to improve the Centre's performance, and worked to communicate the new plan to the Tokyo Centre Governing Board and member CEOs. Tokyo Centre members supported expansion of the Tokyo Centre staff with seconded employees possessing more extensive management experience, technological backgrounds, and better English language capability. Members are also now committed to providing a senior executive to serve as an Exit Representative for each peer review, further supporting improvements in the Tokyo Centre's peer review process.

With these positive steps well under way at Tokyo Centre, WANO turned much of its efforts to completing the 12 projects launched in response to the five recommendations of the Post-Fukushima Commission and looked to integrate processes and practices across the regions. As head of the Oversight Committee, Hawthorne briefed the Governing Board near the end of 2013 on the status of WANO's progress since the Shenzhen BGM. The effort to increase the frequency of corporate peer reviews to every six years was on track, as were plans for a follow- up internal assessment focusing on gaps highlighted in the 2012 assessments. In addition, WANO had successfully developed a severe accident management (SAM) plan through the expansion of emergency preparedness centres in Atlanta and Moscow. An early notification system to alert CEOs of incidents in order to prepare them for possible media enquiries had also begun functioning, Hawthorne reported. Most impressive, perhaps, was the progress the ELT had made on a common process to assess nuclear safety performance, a programme that would eventually lead to ranking plant performance.[39]

Several other projects had been less successful, Hawthorne noted. Establishing design baselines for each reactor type had proved difficult, and the goal of granting equivalency of peer reviews for like-minded organisations, despite there now being a defined process, remained some distance into the future. Moreover, although major strides had been made in improving WANO's visibility, the pilot attempt to issue media releases of peer review summaries under the umbrella of improving transparency had proven flawed and was abandoned, reopening the Governing Board's internal discussion of how best to balance transparency with confidentiality. Ellis explained that "visibility is to promote WANO by describing what we do and what we stand for, and transparency is amongst our members and is what we find." It was an important distinction, and the issue continued to be debated among the governors.[40]

Another lingering concern for the Board of Governors was the implementation of the Nuclear Safety Performance Assessment (NSPA) project, scheduled to be presented at the Toronto BGM in October 2015. The ELT had reached agreement to establish a common WANO process using best practices from the regions. As decided at the Shenzhen BGM, WANO would limit distribution of the results of the assessments to CEOs in a closed session at the BGM. While there had long been opposition to plant rankings, NSPA, as one governor explained, was to gauge a station's overall performance relative to the rest of the industry in relation to established standards of excellence. The necessity for such a presentation to CEOs gained support among the regional boards, and the project, subsequently renamed WANO Assessment, became a priority for the ELT. WANO started conducting WANO Assessments in the autumn of 2014, marking a major cultural shift in the organisation and a watershed moment in its history.[41]

The passing of a quarter of a century since WANO's founding and the considerable changes in the nuclear landscape in the post-Fukushima world made WANO officials keenly aware of the demographic challenge they would face in the future. The phase-out of nuclear power in Germany and elsewhere in Europe and the shutdown of the bulk of the Japanese nuclear fleet, as well as the uncertainties and political controversies regarding restarting the plants, would likely sour young engineers and others on entering the field, leading to a dwindling and ageing workforce, most WANO governors believed. At the same time, new nations were building or planning nuclear plants and required a knowledgeable and trained workforce to operate them. Both Régaldo and Ellis believed a WANO youth initiative was critical in attracting fresh talent into the industry and maintaining a high level of safety culture.[42]

The initiative for a youth movement came from Moscow Centre. To celebrate both the 60th anniversary of the 26 June 1954 commissioning of the first commercial nuclear power plant at Obninsk, the "Science City" southwest of Moscow, and the 25th anniversary of WANO, Moscow Centre brought together young nuclear professionals from all the Centre's member organisations in the Moscow Centre Youth Movement. It was a brilliant idea to honour past achievements while emphasising those who would be the industry's future. To highlight the importance of the gathering for WANO, Régaldo and Ellis participated with Anatoly Kirichenko, the first Deputy Director of Moscow Centre and Sergey Kushnarev, Executive Vice President of the Nuclear Society of Russia. The Governing Board applauded the initiative and urged that the Moscow concept become a WANO-wide movement. Soon after, the Russians drafted an expanded WANO Youth Concept and a WANO Young Professional Policy; those documents became the basis for the WANO Young Generation movement. The intent was to involve the youth group, consisting of those 40 years of age and under, directly with WANO's programmes, with them acting as observers during peer reviews to learn the operational functions of the organisation. The Board also recommended that the younger generation and the growing number of women in the nuclear industry be recognised at the Toronto BGM.[43]

As Moscow Centre was leading the youth movement initiative, events in Ukraine tested WANO's "safety has no borders" axiom. In February 2014, in response to a revolution in Ukraine in which President Viktor Yanukovych was ousted by those favouring closer ties to the European Union, a rebel government was established in the eastern Ukraine region of Donbass. The situation soon led to an armed conflict. The military situation posed a challenge to finding an adequate number of volunteers given the uncertain security confronting WANO peer review teams scheduled to visit plants in south Ukraine. Régaldo worried that the civil strife might isolate the Ukrainian nuclear plants even though they were located some distance from the fighting. WANO had faced security concerns for reviewers before – particularly in Pakistan where the unstable political situation made it difficult to organise a peer review team. The same was true for Iran, where WANO had experienced difficulties assembling review teams to visit the Bushehr site, largely because of visa limitations rather than any danger once in the country. Both challenges had been surmounted. Moscow Centre's Director, Mikhail Chudakov, immediately offered to provide a security plan to assist in obtaining the peer reviewers. In addition, ties between Rosenergoatom and the Ukrainian state energy operator Energoatom remained open, and the Moscow Centre worked unfailingly to provide peer reviewers. The quick response demonstrated that WANO's nuclear safety objectives could overcome political issues between governments. In Régaldo's opinion, "there has been, up to now, no confusion between the political situation and the civil war in the east and the support from Moscow Centre."[44]

Nonetheless, the situation in Ukraine spurred the Governing Board to broaden its view of the industry. Ellis, supported by Governor Robert Willard, the President and CEO of INPO and a representative of Atlanta Centre, suggested that the WANO Governing Board rethink what constituted risk in general rather than looking solely at plant performance. In his mind, it was important for WANO to have a strategic overview of international events and to be aware of factors that might challenge performance.[45]

In another change to adapt to the changing commercial nuclear power environment, the China National Nuclear Corporation (CNNC) proposed that WANO open a centre in Beijing. Chen Hua, a Governor representing Tokyo Centre, explained that since Fukushima, his nation's nuclear safety authority had spent two years conducting a review and, as a result, "the Chinese government has attached great importance to nuclear safety" and "would like international experience to support us." In April 2015, the WANO Strategy, Governance, and Nominating Committee reported to the Governing Board that the idea "may be a good opportunity for WANO" and should be given deep consideration. There were many details to be analysed and hammered out, most hinging on the potential impact of such a move, the first such major regional change in 25 years, particularly for Tokyo Centre, in which the organisation had invested so much in strengthening its activities. All the governors agreed and, given the expected growth of nuclear power elsewhere in Asia and on the Indian sub-continent, asked Ellis to prepare a business case for various options in reorganising WANO, including the creation of a fifth centre in Beijing.[46]

****

Since the New Delhi BGM in 2010, WANO's strategy was to keep adapting to the changing nuclear landscape by pushing for consistent, integrated programmes and more tightly aligned performance among all members. Stricker and Felgate had been preaching the gospel of an integrated WANO since the New Delhi BGM in 2010. The WANO leadership was convinced that the organisation would be best served by a "One WANO" concept consisting of a more unified governance structure, better communications among the regions and a high level of member accountability if its top priority of nuclear safety were to be achieved. The theme was repeated and reinforced by Régaldo and Ellis at a Site Vice Presidents' and Plant Managers' conference in Dusseldorf in September 2014. Ellis reminded the 80 delegates representing 27 countries and areas that although safety cost money and resources, accidents were profoundly more expensive. Amid changes in the industry, continuous improvement of nuclear safety must remain the top priority for all WANO members. To accomplish this, there could be no "silent" plants; stations that did not report operating issues. Shared operating experience drawn from all plants was vital to understand the risks posed in each plant. To emphasise the importance of this function, INPO had made event reporting a requirement for participation in its international programme. "As the people directly accountable for the day-to-day operations of your plants," Ellis told the conferees, "safety begins with you. You must absolutely understand how your organisation identifies, evaluates, and mitigates risk at all levels. Above all, you set the tone for your organisation."[47]

At the Shenzhen BGM, the delegates had made clear that the 12 projects outlined by the Post-Fukushima Commission should be completed by the 2015 BGM. For an organisation that had historically acted with caution, taking care to discuss fully each policy and task, WANO and the Oversight Committee moved remarkably fast toward achieving that mandate. About a year in advance of the BGM, Hawthorne told the Governing Board that the Oversight Committee continued to be impressed with the level of progress being made on the projects. Most were well in hand, such as the regional centre and London Office internal assessments and WANO operational safety assessments and rankings of plants. The Oversight Committee reported that it would highlight the plants that were not meeting their obligations at the closed CEO session at the Toronto BGM. All the centres were scheduled to have an emergency response plan in place by the end of the first quarter of 2015. All agreed that a 72-hour response time for support from WANO was the most realistic scenario. The corporate peer review project to conduct a review of every member by the end of 2017 was also on track, although the periodic frequency of the reviews remained to be determined. While a few projects had been more difficult to accomplish, overall Hawthorne was pleased with what had been done. WANO had challenged itself to complete the post-Fukushima goals, he explained to the Governing Board, and "it is important to realise how far we have come." The governors and ELT agreed that the Oversight Committee had been extremely valuable in shepherding the completion of the Post-Fukushima Commission's recommendations, but that its future role remained undecided.[48]

As the culmination of the work to implement the Post-Fukushima Commission report drew to a close, Ellis decided that the time was ripe to draft a re-envisioned and eye-catching WANO long-term plan for the period 2015–2019. Compass, as the plan was titled, was to simplify and clarify WANO's programmes and direction for the organisation to become the global leader in nuclear safety. Compass was recognition of the evolution of the nuclear world and the need to reinforce WANO and its mission to "maximise the safety and reliability of nuclear power plants worldwide". It was a plan to buttress WANO's traditional activities while preparing for the future. The five-year plan outlined the four strategic challenges "confronting the nuclear industry and the sustained effectiveness of WANO". Not surprisingly, the initial driving force was full implementation of the 12 recommendations made by the Post-Fukushima Commission, "the pathway for WANO to correct identified shortfalls. Full implementation," Régaldo said, "will require going the extra mile. To reach the expected level of performance, WANO will need to review these results with a long-term perspective, enhancing the quality of its programmes and worldwide consistency." WANO's second challenge was to reinforce its support to new entrants and rapidly expanding fleets "so that the highest possible level of nuclear safety can be ensured". The third effort was to increase WANO's appeal to "young talents" and to offer personal, technical, and international development opportunities. Bringing in younger people, Régaldo held, would bring the benefit of "new ideas and visions". The fourth challenge was internal to governance: WANO had to become "increasingly more integrated as an international organisation" to become One WANO.[49]

The Compass plan incorporated the entire nuclear spectrum, a roadmap based on past experience and on the "twists and turns" anticipated in the road ahead, according to Ellis. He saw the plan as "an invaluable resource for the world's fleet of operating reactors". He also believed the plan would be able to "adapt to an industry that will dramatically expand in some parts of the world and slowly phase out in others". For existing and future fleets, WANO would "continue to support and set the standards of high performance... and build and maintain a highly trained, professional workforce". Ellis stressed the necessity of forging a "stronger WANO through consistent, credible products and programmes" through common goals, principles and standards. "Our first commitment is to WANO as a whole, hence independent or autonomous approaches do not improve overall nuclear safety and reliability," he declared. Finally, Compass would help WANO deal effectively with both the beginning and end of the nuclear life cycle by instilling "superior standards among new industry entrants" in areas where nuclear energy had not been a part of the energy supply mix, such as Asia, Africa and the Middle East, while maintaining those standards for older plants in North America and Europe that were approaching extensions, end-of-life, and decommissioning.[50]

As part of its programme to forge a stronger WANO and to identify and support poorly performing plants – those that posed the greatest operational nuclear safety risk as compared with the rest of the industry – WANO initiated the "Plants of Focus" programme in early 2015. A logical extension of plant rankings and the organisation's mission to improve operating safety, Plants of Focus drew on the experience of the regional centres to zero in on at-risk plants and enable regional governing boards to contact the CEO of that utility so that WANO could provide required technical assistance, help develop a formal recovery plan and implement an enhanced monitoring process for each Plant of Focus. The programme was a blend of regional initiatives, with each regional director responsible for implementation of the policy in his region. WANO's London Office developed the guidelines in consultation with the regional centres and, in addition, was responsible for oversight to ensure consistency among the regional centres.[51]

The Compass plan was brief and direct. It was to be a guide, not a detailed work plan, for WANO staff in the regional centres and the London Office, as well as a communications vehicle to share WANO's intentions with all members. It did not try to say or do too much. It was designed as a booklet, to be easily carried and quickly referenced. In that sense, it captured the intent of WANO's "new look" era of communications, intended to reach as many members as possible through print and digital formats, linking members not only to WANO's websites, but also to Facebook, Twitter, YouTube, LinkedIn and Flickr, a reflection of the effort to reach out to younger people in the industry. Compass, WANO officials hoped, would guide "the world's nuclear operators on their path to excellence."[52]

****

The history of WANO has been remarkable, from its creation out of the embers of Chernobyl to its struggle to build an effective international organisation to its recent push to achieve the once lofty, but now realised, goals of its founders. Critical organisational changes in WANO's governance and policies since 2010 have strengthened members' commitment. By naming the managing director of the London Office WANO's chief executive officer in 2014, the Governing Board invested the responsibility of the position with the authority it had lacked. Ellis seized that opportunity, as an equal and full member of the WANO Governing Board, to push through changes long discussed, but only tenuously approached, into reality. Ellis streamlined Governing Board meetings to make them more efficient and less duplicative; he strengthened the staff of the London Office in order to provide more oversight and more direction to WANO's four technical programmes, and added communications as a fifth and vital aspect in the creation of a One WANO. In making these changes, the London Office has improved collaboration with the regions and established its primacy in providing guidance, oversight, direction, and leadership of WANO's core programmes.

Another long-standing WANO issue, the cooperation and coordination of the Executive Leadership Team, had also undergone a transformation. Once a battleground between the regional centres and London, the ELT was fully integrated into WANO's governance and its members were now active participants at Governing Board meetings.

Régaldo and Ellis seized upon Fukushima to build WANO's organisational resources to assess the weaknesses revealed by the accident and lead the members to accept and embrace higher operational standards and more stringent levels of accountability, thereby strengthening the organisational resources and effectiveness throughout all the regions. Moreover, WANO made these changes with a speed few, if any, international organisations could match.

The renewed commitment and cultural shift of WANO members, beginning with the New Delhi BGM and intensifying after Fukushima, enabled the WANO leadership—the Chairman, CEO, Governing Board, and Oversight Committee—to expand their roles in shaping WANO's transformation. That they were able to accomplish not only the projects outlined by the Post-Fukushima Commission, but also able to resolve many of the long-standing issues that WANO had often discussed, though rarely confronted directly in its first 20 years, was a testament to their disciplined approach and ability to build a powerful consensus for change and progress. Nonetheless, it is important to understand that their accomplishments have been built on the work of those who preceded them. Certainly, the steady growth in the role from Director of the Coordinating Centre to Managing Director to Chief Executive Officer of WANO was a crucial change that finally gave the CEO status and authority to match the position's responsibilities. By making the CEO a full voting member of the Governing Board, (a transformation in which Ellis took every advantage to strengthen the London Office and WANO's programmes and governance), the members conceded that the organisation could function more effectively and more efficiently under central guidance rather than as a loose confederation of regions.

The authorisation of central management and oversight in the London Office did not come easily. In expanding London's capabilities, the office had to demonstrate the wisdom of the decision to increase resources and staff in support of WANO's programmes including Peer Reviews, Technical Support Missions, and Performance Indicator data. As the graphics in the central pages demonstrate, WANO has steadily succeeded in improving the operation and performance of its members' nuclear power plants. In addition, by making Communications a core programme in 2013, WANO fostered increased contact with member plants and communicated more effectively, with a more consistent voice, with the membership than in the past, particularly in integrating the regional centres so that they speak to their members with a One WANO voice. For the first time in its history, WANO was intending to create among its members demand for its services rather than simply "pushing" information to them.

****

The promises made at the Moscow BGM in May 2013 to implement the recommendations of the Post-Fukushima Commission by the Toronto BGM in

October 2015 were seen by many as a daunting, if not impossible, task. If the past were any indication, WANO's record over its history for implementing organisation-wide improvements had been patchy, at best. Earlier WANO reviews such as the Franklin-Hall Report or the Kingsley Report had been accepted and then largely ignored due to lack of interest and/or resources on the part of some regional centres. In addition, after Chernobyl, WANO had done a creditable job in managing its internal affairs and avoiding another such accident with major nuclear safety consequences. However, Fukushima, an event driven by external factors, had demonstrated that outside causes also carried serious safety consequences. Effective implementation of the Post-Fukushima Commission recommendations would require a significant move away from the traditional WANO responses of the past. As chair of the Oversight Committee, Hawthorne realised that such a change "would require a mind shift from all of our members and a much stronger commitment of time and resources to those activities" if WANO were to move forward with the Commission's recommendations. "Every member and every region," he said, "had acknowledged that a new WANO had to be created and that they had to make the investment necessary for it to succeed. It is fair to say that a number of the recommendations that got support in a post-Fukushima world had been tried and could not get support in the era before the accident. Fukushima," Hawthorne concluded, "was a crystallising event" that gave new urgency to WANO.[53]

Mitchell agreed. In his mind Fukushima was a "wake-up call". His motto became: "Never let a good crisis go to waste". In an impassioned speech to a "stunned" WANO Governing Board at its April 2011 meeting just after the accident, Mitchell stated that the event put the industry and WANO at a "crossroads". Could WANO become an effective organisation or should the Board "just end this experiment because it's too complicated or too hard having an international organisation made up of 33 different countries and areas that are going to be able to make a difference?" he asked. In his view the industry had reacted slowly to Three Mile Island and Chernobyl. "We stood back and let regulators, politicians, and others act". This time, he said, the industry, led by WANO, "should get out of the blocks early and set the agenda. We should say, 'Yes, this is what happened and this is what we're doing about it'". He wanted the nuclear power industry to act first and demonstrate that it "didn't need to be pushed". As a result of his speech, Stricker selected him to lead the Post-Fukushima Commission and bring back its recommendations on ways to strengthen WANO within six months.[54]

The urgency of WANO's response was a sharp departure from its past. "We did not start with a tabula rasa," Mitchell later said of the Commission. "We took the work approved in Delhi in 2010, such as increased staffing, greater frequency of peer reviews and the like, but there just wasn't the oomph behind it. We were able to seize on that, add some very good ideas, and put a sense of urgency behind it". The ongoing audits and follow-up assessments of the regional centres and the London Office, led by Team Leader Matt Sykes and his successor G. Wayne Robbins, were crucial to changing WANO's approach. "The assessments," Mitchell believed, "turned out to be so powerful because they were fact-based, not emotional, pointing out the gaps between what the regions said they were going to do and what they were actually doing — whether that meant resources or programmes or support. For the first time we had an objective view of where each region was relative to WANO's objectives, followed by a strong focus from the Oversight Committee on closing those gaps and getting things done". The assessments and the follow-up on the Post-Fukushima Commission's recommendations in Mitchell's view "turned out to be a very big accelerant in the change."[55]

At the Toronto BGM in October 2015, almost four years to the day after the Post-Fukushima Commission's recommendations had been ratified at Shenzhen, WANO

CEO Ken Ellis reported, as promised at the Moscow BGM two years before, that all Post-Fukushima measures had been implemented or fully launched. Ten of the twelve projects stemming from the Commission Report had been completed and significant progress had been made on the other two—Design Safety Fundamentals, not expected to be fully implemented for several years, and the Emergency Support Plan, which had an anticipated completion date in the spring of 2016. For many WANO veterans, the most important accomplishment had been the establishment a worldwide WANO Assessment Policy that would lead to comparative rankings of nuclear plant operations. With this policy in place, WANO could better identify and assist those plants "that represented a higher operational nuclear safety risk compared with the rest of the industry". The Plants of Focus initiative, Ellis said, would be fully in place by March 2016. Demographics was another area meriting WANO's close attention, Ellis stressed. "The nuclear industry is at a demographic crossroads: some parts of the world are experiencing an exodus of experienced workers, while others face a skills shortage. WANO must assist the transfer of knowledge and assistance between the two sides of this crossroads," through a WANO-wide Young Generation programme. WANO, too, must change to keep up with the shifting nuclear landscape, to adapt its programmes to meet future challenges brought on by new plant building, plant extensions and plants at the end of their lifetimes. WANO would strengthen its cooperation and alignment with the IAEA in working with countries first entering the nuclear power community, Ellis stated. To meet such changes and challenges, Ellis said, members in the future would no doubt consider "the need for reorganising WANO, including the addition of a possible fifth centre."[56]

Ellis' presentation left no doubt that WANO had changed dramatically since 2011. While his graphs of performance indicators [included in central insert] demonstrated industry improvement across the board, the chart on WANO staffing between 2011 and 2015 displayed the growth and maturation of the organisation. Total staffing had increased from 139, rising nearly twofold by 2015. Over that period, Moscow Centre staff grew from 17 to 79, Paris Centre from 50 to 125, and Tokyo Centre from 27 to 74. The London Office nearly doubled in staff, attracting a greater number of secondees in the middle of promising careers rather than those soon scheduled for retirement as had often been the case in the past. The influx of new staff enabled the London Office to be more collaborative with the Executive Leadership Team and with the regions at the programme level. "Thanks to Ellis," Régaldo said, WANO has "made great improvements in enhancing our internal skills, strengthening our operational programmes and developing collective and transversal teamwork among our regional centres". Robert Willard, the CEO of INPO echoed Régaldo, whom he also acknowledged for instituting improvements at WANO. "Much of the credit goes to Ellis who was the real energy behind many of the London Office's initiatives and the ability to pull the regions together to achieve the progress necessary for WANO to advance. He flat-out took personal ownership of some of the flagging Post-Fukushima initiatives and saw them through to a successful completion."[57]

WANO's successful implementation of the Post-Fukushima Commission's recommendations could not have occurred without what Mitchell termed as a "cultural shift" among WANO members, in the sense that WANO had become "an organisation that is going to monitor plant performance and is going to hold its members accountable". The first step in moving toward the shift, he believed, was Stricker's initiative to engage utility CEOs throughout all the WANO regions, but particularly on the Tokyo Centre Governing Board. Their involvement after Fukushima and WANO's assistance in strengthening Tokyo Centre were essential for WANO moving forward. Although the accident gave urgency to reassessing and reforming WANO based on the governance changes approved at New Delhi and the Post-Fukushima Commission recommendations ratified at Shenzhen, individuals had to take ownership to ensure their success. Mitchell first credited Felgate, who turned the Post-Fukushima Commission's four core recommendations into 12 specific projects, making them easier for WANO members to understand and support. He gave highest marks to Ellis, who, he said, "stepped in and stepped up. He brought a certain energy and focus to the issues [raised by the Post-Fukushima Commission]. He saw how important this was. He galvanised the ELT. The Governing Board tried to help by making him a CEO and voting member of the Board". Mitchell believed the change Ellis made in sending issues to the regional governing boards for comments and suggestions to the ELT before being considered by the WANO Governing Board was brilliant. "When the main governing board now says yes [to a policy] and goes back to the regions to get it implemented, they're on side. It was a very simple change that fundamentally altered the dynamic of how the organisation functioned". The cultural shift and the acceptance of broad changes in WANO succeeded, Mitchell concluded, because it was "an international effort."[58]

Willard, too, recognised the importance of WANO's cultural shift. "WANO evolved from a great deal of focus on itself and its own challenges—trying to rationalise where it was deficient as a broad organisation and in governance, and began to shift its focus to the international industry and began to talk about industry performance in very different ways than it had before. There has been a paradigm shift from an inward focus of WANO on itself and its image to a more mature look into the international industry it was there to serve in an effort to understand the industry in context and recognise where the risks are and are not internationally."[59]

Hawthorne agreed with Mitchell and Willard regarding the importance of fully staffing the regional centres and the London Office and the need for a stronger role for the London Office; for it to receive more resources and have more authority, to become more of a leader than a coordinator and consensus-builder. Making the Managing Director the WANO CEO was a crucial step in that strategy as it created a position "capable of taking action on its own to help deliver the plan and with the necessary control of resources to actually implement things". He noted that in many ways the Stricker/Felgate and the Régaldo/Ellis teams were much alike. Both chairmen were active, full-time leaders who travelled tirelessly in support of WANO. He recognised Ellis as a very good organiser with the skillset "that WANO needed". Hawthorne noted that the completion of the Post-Fukushima Commission recommendations was crucial to WANO's future. As the organisation strengthened its programmes and overall resource pool, he explained, WANO will be "much more resilient to changes in personnel. Leadership has been enhanced because there's more structure behind them". He pointed to the work that Stricker, Felgate, Régaldo, and Ellis had done with the IAEA in "convincing people in the international community" of the importance and value of WANO. "They have given WANO a more external persona that in the past. As we build programmatic strength, we will better define authority and accountability, and that should allow us to make changes in individuals without missing a beat. We're all getting older and so whatever we're doing now has to be built to last."[60]

****

More than a quarter of a century ago, in the aftermath of Chernobyl, WANO was created with the vision of becoming a special international organisation to establish operating standards beyond those of state regulators in order to improve the safety and reliability of the entire commercial nuclear power industry. The idea of self-regulation was not new, but WANO was the first group to work exclusively at an international level. Jacques Régaldo observed that "more than ever before, WANO was perceived as a 'self-regulator' for the community of nuclear operators and had a much stronger image" as a result. The industry had experienced enormous economic, environmental, geopolitical and technological changes during that 25-year period. So had WANO. While its mission remained steady—the continuous enhancement of the safety of all commercial nuclear reactors—the WANO of 2015 bore but superficial resemblance to that young organisation of 1989. It had learned well the lessons of history. WANO's resources had grown, its staff had expanded, its members were more deeply committed, its strategies more sophisticated, its governance more unified, its programmes more closely aligned and its self-regulatory function more authoritative.[61]

An association created out of Chernobyl amidst the competing ideologies of the Cold War, WANO leveraged a second crisis years later not only to reinforce nuclear safety, but to solidify organisational trust and commitment among its members. "From a functional point of view," Régaldo told members in Toronto, "we have reinforced our organisation." Under Ellis' initiative, he said, "the Executive Leadership Team has strengthened the consistency of our programmes and ensured their implementation" in each region. Moreover, the London Office was "now fully able to play its role in overseeing and coordinating the implementation of [WANO's] decisions and programmes to achieve greater consistency worldwide". Fukushima, rather than destroying WANO, left the organisation and members stronger and more committed to higher standards and more stringent accountability – and with greater resources. The accident forcefully demonstrated that 25 years of hard, sometimes frustrating, work had created an association unparalleled in its ability to adapt to changing conditions and influence industry performance across national, cultural and linguistic barriers. WANO had raised its visibility internationally, without compromising its internal transparency and the confidentiality of its members. Perhaps WANO's greatest accomplishment over its first quarter of a century was to achieve both internal and international credibility. Out of many WANOs, One WANO had emerged. [62]

After reflecting on WANO's past, Régaldo mused one morning about his vision of WANO's future. "Our main role is to prepare for the future," he maintained. Initially, WANO was "a confederation of regions" with a small coordinating office. But after Fukushima, he said, WANO underwent a paradigm shift. In this "new era" WANO needs "a more integrated organisation, with competent, fully staffed regional offices," he said. "For me, the future is an integrated company with strong regional resources centres connected on a daily basis to their members. In the future we have to welcome new ideas from young talents and other industries." He also saw great value in creating an advisory or expert committee to advise the chairman and WANO Governing Board. Experts from aviation, chemical, or oil industries that also dealt with safety and risk management could bring a fresh perspective and challenge WANO to ensure that it continually improved. The advisory committee would be a valuable watchdog, Régaldo believed. "We will always have to question ourselves," he concluded, and peer reviews had taught the value of outside observation.[63]

Much had been learned over WANO's first quarter century. Three Mile Island, although having limited impact outside North America, demonstrated that a unifying event could break down walls and end isolation among commercial nuclear power companies. Chernobyl demonstrated that a major accident could also break down international barriers between countries. And Fukushima proved that long-standing cultural and geopolitical differences might also be resolved in the interests of global nuclear safety. From a fledgling association trying its wings, in 25 years WANO had evolved into an organisation with global credibility, authority and recognition – a "more mature organisation with clear objectives and higher level of requirements and commitments supported by stronger, more effective programmes", Régaldo asserted. This successful persistence for safety could not have occurred without the unflagging commitment of industry executives who, over that time, came to appreciate the fundamental lesson of organisational strength based on a commonality of interests among all nuclear utilities. In addition, those executives came to embrace and value WANO's programmes and the operational improvements they provided. It was critical to the industry's future that those leaders continue to make WANO's values an integral part of their corporation's safety culture. Moreover, WANO's achievements often occurred under difficult political, economic, environmental and security conditions. Members, nonetheless, were able to overcome their differences and put aside self-interests to focus on nuclear safety. WANO's ability to place foremost the common good, Régaldo held, demonstrated that "nuclear safety transcends borders and has become One WANO, not four or five". WANO stands as a testament to those individuals worldwide who believe that compromise and complacency have no place in nuclear safety and that collaboration and commitment to continuous improvement can secure a safe and strong industry across all borders. [64]

For nearly two decades many WANO members supported their association's goals, but did not fully commit the resources to achieve them. Long-time WANO veterans are keenly aware of the cultural shift that occurred among WANO members since Fukushima. One recalled the days when WANO "operated more like a club. You joined and you signed a charter, and went home. Now we understand that WANO is about improving performance of the plants and being able to hold ourselves accountable to each other for our performance. I am in awe that WANO has developed a common language and approach from people from different parts of the world, different companies, speaking with a degree of candour. It has been a long time coming, but achieving that on an international scale, with 33 countries and areas, has been the most exhilarating experience of my life". Duncan Hawthorne also saw a new WANO built on the shoulders of its past. After the Fukushima accident, Hawthorne said, the members "stepped up" and confirmed "the importance of the role of WANO. I have seen a greater level of commitment from our members in the past two years that I have seen in the previous eight while serving on the WANO Board. There is no question that WANO today," he maintained, "is much better equipped, much stronger, much more resilient, much more self-critical and challenging of itself that it has been since its original formation."[65]

Régaldo echoed the importance of the historical evolution of WANO. Created near the end of the Cold War, WANO began as a federation of regions, he remarked, "and any other kind of structure would certainly not have been possible. But to ensure we remain credible today," he told delegates in his closing remarks at the Toronto BGM in October 2015, "WANO must become more integrated and develop a more international approach, to find a balance between the role of the regions, which ensure resources and close relations with the members, and the overriding need for greater consistence across the global organisation, to be One WANO. In 26 years, we have been able to demonstrate our ability to work together and make progress by involving numerous organisations from various countries and cultures," he said. "We have been able to show our independence, our commitment and our will to enhance nuclear safety above any other consideration. This is our strength and this, I believe, should be our common pride."[66]

****

When first formed in 1989, WANO's international structure was somewhat uncertain, the mission and goals of the organisation held together by the genius of Lord Walter Marshall. Succeeding WANO chairmen, WANO presidents, the WANO and regional centre governing boards, the managing directors and CEO of the London Office, senior utility executives, and an array of self-assessments worked toward the gradual improvement of programmes, expanded resources, and altered the basic governance structure, thereby strengthening the organisational core, producing greater accountability, and giving WANO long-term stability.

Over nearly a quarter of a century the gains were incremental, culminating in the governance changes ratified in Delhi in 2010. Fukushima and Shenzhen, however, provided the tipping point for WANO to find a balance between the regions and the nucleus, to emerge as a more unified, more stable, more credible, and more mature organisation. The challenge in the future for WANO will be the challenge it has faced in the past: to avoid becoming complacent. Drawing on WANO's history, its members have built toward the progress reported at Toronto by embracing the cultural shift that took place after Fukushima and applying the wider lessons learned from that event to raise nuclear safety operations at each member unit. The gains cited at Toronto confirmed that continued vigilance and commitment can ensure a strong and stable WANO.

WANO has a special history, and that history should not be taken for granted. The responsibilities for improving the safe operation of commercial nuclear power that WANO embodies have enormous consequences for the well-being of our world. The mantle of safety is never worn lightly. What WANO has accomplished over its history is not rare – it is unique among global industries. No other group of international concerns has progressed so far in creating a self-regulating structure among all its peers. No other industry invites peer teams into its facilities, opens up its programmes and processes to close scrutiny, bases operational changes on the recommendations of potential competitors, and upholds a joint commitment to work toward best practices and excellence above and beyond government regulations and standards, all in the name of the common good. George Felgate once compared WANO's concept of self-regulation to a rough jewel. The gemstone required constant polishing and buffing to smooth off its unfinished edges. The imperfections did not diminish the value of the stone, but the continued burnishing added lustre, refining and enriching the stone's utility and purpose.[67]

And so it continues. After more than a quarter of a century, WANO remains a work in progress, constantly strengthening its governance, its programmes and its operational reach. WANO has changed as the industry has changed. It has adapted to and adopted new technologies. It has become an effective international organisation that makes a difference. And WANO has begun a transition to a new generation of nuclear plant operators who will write the next chapters of its history.

ENDNOTES

PREFACE

1   Figueiredo quoted in "Improving the Efficiency of WANO Paris Centre," *Inside WANO*, Vol. 22, No. 4 (2014). Figueiredo received a Nuclear Excellence Award in 2003.

2   *Washington Post*, January 3, 2015.

1 FORGING A GLOBAL SAFETY NET

1   Anatoly Kirichenko Oral History, 17 May, 2013, 8; Stanley J. Anderson Oral History, 20 September 2010, 16; Zack T. Pate Oral History, 5 March 2013, 1-3; author's conversation with George Hutcherson, 5 March 2013; "Nuclear Glasnost," Nuclear Industry (2nd Q, 1989), 3.

2   Biography of Nicolay F. Lukonin, WANO Special Award Program, 21 May 2013; The Lord Marshall of Goring Biography, WANO, 1990; Andrew Clarke Oral History, 11 May 2013, 18; Kirichenko Oral History, 17 May 2013, 7; Clarke Oral History, 11 May 2013, 64; Saitcevsky in Proceedings, World Association of Nuclear Operators, Inaugural Meeting, 14-15 May 1989, Moscow, USSR, Inaugural Meeting, 19.

3   Proceedings, World Association of Nuclear Operators, Inaugural Meeting, 14-20 May 1989, 4-5, 54; Rhona Tehrani, WANO Fact Sheet on Inaugural Meeting, 13 July 1989; see also Zack T. Pate Oral History, 24 January 2013, 19.

4   Zack T. Pate Oral History, 24 January 2013, 19; Proceedings, World Association of Nuclear Operators, Inaugural Meeting, 14-20 May 1989, 4-5, 54; Rhona Tehrani, WANO Fact Sheet on Inaugural Meeting, 13 July 1989; see also Zack T. Pate Oral History, 24 January 2013, 19; "Nuclear Glasnost," Nuclear Industry (2nd Q, 1989), 3; William S. Lee III, notes from the WANO Inaugural Meeting, 15 May 1989, WANO History Project files. Of the participants, only the People's Republic of China did not sign.

5   Gorbachev quoted in Mark Joseph Stern, "Did Chernobyl Cause the Soviet Union to Explode?" *Slate*, 25 January 2013; James Graham, "Gorbachev's Glasnost," HistoryOrb. com.

6   Marshall speech in Proceedings, World Association of Nuclear Operators, Inaugural Meeting, 14-20 May 1989, 34; Clarke Oral History, 11 May 2013, 4-6.

7 Ibid., 4.

8 Philip L. Cantelon and J. Samuel Walker, *Core of Excellence: A History of the Institute of Nuclear Power Operations* (Rockville, MD: 2012), 59-60, 117-118, hereinafter cited as Cantelon and Walker, *Core of Excellence*. A former admiral, Stanley J. Anderson, headed the international programme. See ibid., 117.

9 Clarke Oral History, 11 May 2013, 11-12, 42; Zack T. Pate Oral History, 24 January 2013, 7. INPO used the term "participant" for international utilities to differentiate them from INPO members who represented utilities holding US nuclear operating or construction licences. The International Participant Program operated so that its activities would integrate with INPO's programmes. Participants loaned staff to INPO, hosted technical conferences and provided operating experience. George Hutcherson to author, 10 February 2014.

10 Clarke Oral History, 11 May 2013, 11-12, 42.

11 Ibid.; Zack T. Pate Oral History, 24 January 2013, 7; "Background: History of WANO," 13 July 1989, 1.

12 Interview with Jacques Leclercq and Rémy Carle, 14 May 2013, 6-8.

13 Ibid. At the time, 8 of 10 light bulbs in France were lit by atomic energy. See remarks of Pierre Delaporte, International Nuclear Utility Executive meeting, 5-6 October 1987, 2, hereinafter cited as Proceedings, Paris, October 1987; Clarke Oral History, 11 May 2013, 13.

14 See George Hutcherson Oral History, 23-24 April 2013, 10-11; Clarke Oral History, 11 May 2013, 28.

15 Clarke Oral History, 11 May 2013, 42; Hutcherson Oral History, 23-24 April 2013, 2, 8-12, 16.

16 Ibid., 23-24; Thomas Eckered Oral History, 25 March 2013, 6-7.

17 Ibid., 4-8; Clarke Oral History, 11 May 2013, 12-14; Hutcherson Oral History, 13. In English, UNIPEDE translates as the International Union of Producers and Distributors of Electrical Energy.

18 Clarke Oral History, 11 May 2013, 12-14.

19 Hutcherson Oral History, 13; Clarke Oral History, 11 May 2013, 12-14, 28.

20 Ibid., 3, 14.

21 "Chronology of INPO/USSR Interactions," 25 June 1990, WANO Archives, Atlanta; Clarke Oral History, 14-15, 27-28.

22 Clarke Oral History, 27-28; Speakers list, programme for International Utility Executive Meeting, Paris, 4-5 October 1987; see also Hutcherson Oral History, 23-24 April 2013, 19.

23 Clarke Oral History, 11 May 2013, 15-18.

24 Ibid., 28; Lord Marshall, Proceedings, October 1987, 3.

25 Clarke Oral History, 11 May 2013, 28-31; Hutcherson Oral History, 23-24 April 2013, 32; Lukonin interview with Vera Lukyanova, 21 March 2013, Lukonin Folder, WANO History Project Files.

26 Clarke Oral History, 11 May 2013, 28-31.

27 Lord Marshall, Proceedings, October 1987, 91, 3-4, 8.

28 Ibid., 6; "Background History of WANO," 2. For a recap of what INPO believed was important for the new organisation, see Pate speech, Proceedings, October 1987, 80-82.

29 Clarke Oral History, 11 May 2013, 16. In Asia the Japanese had the longest experience with commercial nuclear power, and Japan would influence the nuclear culture of Asia for more than two decades. After Fukushima, the Chinese and Indians occupied more central roles in redefining the approach to nuclear safety.

30 Ibid., 14-17; Lord Marshall, Proceedings, October 1987, 6.

31 Ibid., 3.

32 Ibid., 6.

33 Ibid., 79, 78.

34 Clarke Oral History, 11 May 2013.

35 Lukonin speech, Proceedings, October 1987, 93; Marshall Speech, ibid., 94.

36 Matsutani speech, ibid., 96.

37 Lukonin, ibid., 118.; Pate, ibid., 82-83; Marshall, ibid., 83.

38 Ibid., 108, 6; Leclercq, ibid., 84, 116; Marshall, ibid., 92; Lukonin, ibid., 93; Georgi Ditchev

speech, ibid., 100; Marshall, ibid., 110.

39 Marshall Speech, ibid., 108-110; Reiner Lehmann speech, ibid., 103-104.

40 Ibid.

41 Lukonin speech at 13th WANO Biennial General Meeting, 22 May 2013; Lukonin interview with Vera Lukyanova, 21 March 2013, Lukonin Folder, WANO History Project Files.

42 Hutcherson Oral History, 23-24 April 2013; Clarke Oral History, 11 May 2013, 46-48, 51.

43 Eckered Oral History, 25 March 2013, 9-10; Hutcherson Oral History, 23-24 April 2013; "Outline for NRC Briefing, 18 May 1988, on World Association of Nuclear Operators," INPO draft, 11 May 1988, WANO Archives, Atlanta, hereinafter cited as NRC Briefing.

44 Clarke Oral History, 11 May 2013, 46-48, 51; Eckered Oral History, 25 March 2013, 5-6, 15. Hutcherson Oral History, 23-24 April 2013, 34-35. The group also discussed how WANO sounded like guano, but decided that "they could live with that." See ibid., 35.

45 "The Formation of the Regional Centres". Section 3.6, untitled report, n.d., 1-3, WANO Archives, Atlanta; "NRC Briefing," WANO Archives, Atlanta. On Prushinsky see Grigori Medvedev, *The Truth about Chernobyl* (London: 1991), 136.

46 Cantelon and Walker, *Core of Excellence*, 124; Eckered Oral History, 25 March 2013, 12-13; Hutcherson Oral History, 23-24 April 2013, 28-29.

47 Hutcherson Oral History, 23-24 April 2013, 30-32. As of this writing, the internal WANO website continues to run off the INPO computer in Atlanta.

48 Ibid., 35-36; 54.

49 Ibid., 37-40. Although the terminology has changed over the years, as of this writing a WANO event is based on the same criteria as an INPO significant event.

50 Ibid., 70-73.

51 Proceedings, World Association of Nuclear Operators, Inaugural Meeting, May 14-15, 1989, Moscow, USSR, 36-39; hereinafter cited as Proceedings, WANO, Inaugural Meeting.

52 Meeting Programme in ibid., 3-9; Report on Moscow Meeting to INPO Employees by Stan Anderson, 24 May 1989, 3-5.

53 Proceedings, WANO, Inaugural Meeting, 87-89; Hutcherson to author, 10 February 2014.

54 Proceedings, WANO, Inaugural Meeting, 89-91.

2 THE SECOND MARSHALL PLAN

1 Obituary, Lord Marshall of Goring, *Herald Scotland*, 29 February 1996; John Baker, Obituary, Lord Marshall of Goring, *The Independent*, 26 February 1996; Marshall's marriage to Lady Ann would last 41 years until his death in 1996.

2 Obituary, Lord Marshall of Goring, *Herald Scotland*, 29 February 1996; John Baker, Obituary, Lord Marshall of Goring, *The Independent*, 26 February 1996; D. Fishlock and L. E. J. Roberts, *Biographical Memoirs of Fellows of the Royal Society*, vol. 44 (Nov. 1998), 299;

hereinafter cited as Fishlock and Roberts, *Biographical Memoirs*.

3 Obituary, Lord Marshall of Goring, *Herald Scotland*, 29 February 1996; John Baker, Obituary, Lord Marshall of Goring, *The Independent*, 26 February 1996; Fishlock and Roberts, *Biographical Memoirs*, 303. Marshall is not to be confused with the character names Lord Goring in the Oscar Wilde play *An Ideal Husband*, written in the 1890s and named for the village.

4 *The Times*, 23 February 1996; John Baker,

Obituary, Lord Marshall of Goring, *The Independent*, 26 February 1996; Fishlock and Roberts, *Biographical Memoirs*, 307.

5   Ibid., 307-309.

6   Ibid., 307-309. See also Minutes of WANO Governing Board, 4 April 1990, 2. After considerable negotiations, Marshall's severance was approximately £250,000. See Fishlock and Roberts, *Biographical Memoirs*, 309 and Clarke Oral History, 11 May 2013, 24.

7   Robert Franklin, "The Challenge," speech to WANO Atlanta Centre (AC) Inaugural Meeting, 23 March 1989, WANO AC files.

8   Lord Marshall of Goring, "The Plan for WANO," 15 May 1989, 2, 8, 10, WC-AC Archives.

9   *Nuclear Industry*, First Quarter 1990, 7; Fishlock and Roberts, *Biographical Memoirs*, 309.

10 Thomas Eckered to the WANO Governing Board, 15 May 1991, WANO Archives.

11 Thomas Wellock, "Separating Technology from Politics: The Fall of the Soviet Bloc and Nuclear Reactor Safety," 11-12 (2012); WANO Archives; Alice Clamp, "Working Without a Net," *Nuclear Industry*, Third Quarter 1992, 10.

12 Quotes from Wellock, "Separating Technology from Politics: The Fall of the Soviet Bloc and Nuclear Reactor Safety," 12 (2012), WANO Archives.

13 Minutes, WANO Board of Governors Meeting, 17 July 1990, 4, WANO Archives.

14 Wellock, "Separating Technology from Politics: The Fall of the Soviet Bloc and Nuclear Reactor Safety," 12-14 (2012); WANO Archives. Although Churchill's Iron Curtain ran from the Baltic to the Adriatic, the Krško Nuclear Power Plant in Slovenia, part of the former Yugoslavia, near the Adriatic was built for Slovenia and Croatia by Westinghouse, not the Soviets. See World Nuclear Association, "Nuclear Power in Slovenia," www.world-nuclear.org.

15 See Jacques Leclercq, *The Nuclear Age* (Paris, 1986), 86.

16 *Nuclear Industry*, First Quarter 1990, 7.

17 *Nuclear Industry*, Second Quarter, 1989, 3; Speech of Dr. Werner Hlubek at Paris Centre Annual General Meeting, 27 March 1990, *passim*; Thomas Eckered to Zack T. Pate, 30 March 1990. See also Jacques Burtheret, "VVER Special Project Status," 12 April 1991, WANO Archives.

18 Eckered to WANO Governing Board, 15 May 1991, WANO Archives.

19 Eckered to WANO President and WANO Governing Board, 12 September 1990.

20 Jacques Burtheret, "VVER Special Project Status," 12 April 1991, WANO Archives; Eckered to WANO President and WANO Governing Board, 12 September 1990; draft letter, Pate to Marshall, 15 August 1991, WANO Archives; Resolution, WANO Atlanta Centre Governing Board, 10 July 1991, WANO Archives.

21 Minutes, WANO Governing Board Meeting, 17 July 1990, 4-5, ibid.

22 Jacques Burtheret, "VVER Special Project Status," 12 April 1991, WANO Archives.

23 Draft letter, Pate to Marshall, 15 August 1991, WANO Archives; Resolution, WANO Atlanta Centre Governing Board, 10 July 1991, WANO Archives.

24 Minutes of the VVER Western Advisory Group Meeting, 20 June 1991, WANO Archives; Memorandum, WANO Contribution to Eastern Europe Nuclear Power Plants, 2 September 1991, WANO Archives; Thomas Eckered to the WANO Governing Board, 15 May 1991, WANO Archives; Minutes of the VVER Special Project Steering Committee, 17 July 1991, WANO Archives; Clamp, "Working Without A Net," *Nuclear Industry*, Third Quarter 1992, 12.

25 Minutes, WANO Governing Board Meeting, 17 July 1990, 5-6 and Attachment 2, Ibid.

26 Ibid. Minutes, WANO Governing Board, 16 November 1990, 3.

27 WANO, Annual Review for the Period Ending April 1992, 3; see Aureliu Laca to Stanley J. Anderson, 2 October 1991 and R. I. Lindsay to Stanley J. Anderson, 7 November 1991, WANO Archives. At the Atlanta BGM the Guangdong Nuclear Power Joint Venture Company and the Chernobyl NPP signed the WANO Charter.

28 Ibid. In 2002, Nasu, then an adviser to TEPCO, a position reserved for retired senior executives, was caught up in a major safety scandal at Japan's largest utility when the company failed to report accurately cracks in its nuclear reactors in the late 1980s and 1990s. TEPCO was suspected of falsifying 29 cases of safety repair records. The company, one Japanese government official stated, had "betrayed the public's trust over nuclear energy." Nasu and several other top TEPCO officials were forced to resign over the incident. See "Heavy Fallout from Japan Nuclear Scandal," CNN.com, 2 September 2002.

29 Author's conversation with Zack and Bettye Pate, 25 September 2013; Programme, WANO Biennial General Meeting, 21-23 April 1991, *passim.*

30 Pate to Marshall, 13 May 1991, WANO Archives.

31 Draft letter, Pate to Marshall, 15 August 1991, 2-11, WANO Archives.

32 See Wellock, "Separating Technology from Politics: The Fall of the Soviet Bloc and Nuclear Reactor Safety," 22-23, WANO Archives.

33 Memorandum, Lord Marshall to Chairmen of the WANO Regional Centres, 19 July 1991, Attachment 1, WANO Archives; WANO, *Biennial Review, 1991-1993*, 15.

34 WANO, *Biennial Review, 1991-1993*, 14.

35 WANO, *Biennial Review, 1991-1993*, 14; Minutes, WANO Governing Board, 16 November 1990, 7; author's conversation with Wesley von Schack, 16 February 2013; author's conversation with Zack T. Pate, 24 December 2013.

36 Minutes, WANO Governing Board, 16 November 1990, 7-8; WANO Chronology, 12; Hutcherson Oral History, 23-24 April 2013, 84.

37 Marshall to Members of WANO, 5 December 1991, WANO Archives; see also Hutcherson Oral History, 23-24 April 2013, 83-84.

38 "Desired Outcome, Pilot Peer Reviews," memo, W. J. Coakley to S. J. Anderson, 30 July 1991, WANO Archives.

39 Author's conversation with Tim Martin, 26 September 2013; Memorandum, Thomas Eckered to Regional Centre Directors, WANO-IAEA Meeting in Vienna, 5 September 1991, 2, WANO Archives; WANO Chronology, 13-15. See also Minutes, Governing Board, 13 November 1992, 6.

40 WANO, *Biennial Review, 1991-1993*, 6-13.

41 WANO, *Biennial Review, 1991-1993*, 6-7; Minutes, WANO Governing Board, 18 April 1993, 4,

42 WANO, *Biennial Review, 1991-1993*, 2; Minutes, WANO Governing Board, 13 November 1992, 4, WANO Archives.

43 Minutes, WANO Governing Board, 18 April 1993, 4-5; ibid. 9 July 1993,4.

44 Minutes, WANO Governing Board, Private Session, 13 November 1992, 1-2.

45 Ibid., 18 April 1993, 2-3, 8.

46 WANO, *Biennial Review, 1991-1993*, 2; Minutes, Governing Board, 13 November 1992, 9.

47 WANO, *Biennial Review, 1991-1993*, 2.

48 Ibid., 3.

49 Resolution of the WANO Board of Governors, 18 April 1993, WANO Archives.

3 PROTECTING THE CORE

1   WANO, *Biennial Review*, 1993-1995, 5; Zack T. Pate to author, 13 February 2014.

2   Andrew Clarke Oral History, 11 May 2013, 79-80; Pate to author, 13 February 2014.

3   Ibid.

4   "Carle: An International Nuclear Leader," Nuclear News, January 1997, 36; Carle to author, 5 March 2014.

5   Programme, International Nuclear Utility Executive Meeting, 5-6 October 1987, n.p; Pate Oral History, 4 March 2013, 30; "Carle: An International Nuclear Leader," Nuclear News, January 1997, 36; Carle to author, 5 March 2014.

6   UTC-infos, 10 January 2002, 4; author's conversation with Zack T. Pate, 29 January 2014; WANO, *Biennial Review*, 1993-1995, 5; Clarke Oral History, 11 May 2013, 84; Carle to author, 5 March 2014.

7   Minutes, WANO Governing Board, 8 July 1993, 9.

8   Ibid., 18 November 1993, 3; ibid., 21 April 1994, 4-5; Clarke Oral History, 11 May 2013, 84-85.

9   "Reliving Apartheid's Dying Days," Johannesburg Mail and Guardian, 5 April 2012; Donald McRae, "My Dad's Soweto Secret," The Guardian, 6 April 2012. Ian McRae's role in electrifying South Africa is based on reviews of Under Our Skin: A White Family's Journey Through South Africa's Darkest Years by his son, Donald.

10 Ibid., author's conversation with Zack T. Pate, 29 January 2014.

11 *Inside WANO*, No. 2 (February 1994), 2; Minutes, WANO Governing Board, 21 April 1994, 5-6.

12 Minutes, Strategic Review Meeting, 21-22 April 1994, 2.

13 WANO, Annual Review, April 1994, 2.

14 George Felgate to author, 25 September 2014.

15 Ibid., 2-4.

16 Minutes, WANO Governing Board Meeting, 20 July 1994, 7-8; Felgate to author, 25 September 2014.

17 Minutes, Strategic Review Meeting, 21-22 April 1994, 5-6, 11.

18 Ibid., 6-10.

19 *Inside WANO*, No. 2 (February 1994), 2 and ibid., No. 5 (November 1994), 6.

20 Minutes, Strategic Review Meeting, 21-22 April 1994, 13-14; Inside WANO, No. 2 (February 1994), 2 and ibid., No. 5 (November 1994), 6; Minutes, WANO Governing Board Meeting, 17 November 1994, 5, 15; Minutes, WANO Governing Board Meeting, 6 November 1995, 6.

21 Minutes, Strategic Review Meeting, 21-22 April 1994, 9-10; Minutes, WANO Governing Board Meeting, 20 July 1995. Saraev, a highly decorated veteran of the Soviet and Russian nuclear power programmes, later became the president of WANO (2004-2006) and the director general of Concern Rosenergoatom. See http://zoominfo.com/p/Oleg-Sarayev188267637.

22 Minutes, Strategic Review Meeting, 21-22 April 1994, 11-13, 16. After the meeting EDF and Koeberg signed a mutual assistance accord, which served as a model for cooperation between other prospective partners. See Minutes, WANO Governing Board Meeting, 17 November 1994, 3.

23 Ibid., 4-6; Felgate to author, 25 September 2014; *Inside WANO*, No. 4 (August 1994), 2.

24 WANO Governing Board Meeting, 17 November 1994, 11-12; Ibid., Closed Session, 2; WANO, *Biennial Review*, 1993-1995, 19.

25 WANO, *Biennial Review*, 1993-1995, 4-6.

26 Ibid., 8-11; Atlanta Centre, Annual Report, 1995, 5.

27 WANO, *Biennial Review*, 1993-1995, 12-13; Minutes, WANO Governing Board, 23 April 1995, 11; ibid., 20 July 1995, 17; ibid., 6 November 1995, 3; Atlanta Centre, Annual Report, 1995, 6-7; ibid., 1996, 7; ibid., 1997, 7.

28 WANO, *Biennial Review*, 1993-1995, 14.

29 Ibid.; Rémy Carle to author, 13 March 2014; Atlanta Centre, Annual Report, 1995, 3.

30 WANO, *Biennial Review*, 1993-1995, 15; Fel-gate to author, 25 September 2014.

31 Minutes, WANO Governing Board Meeting, 20 July 1995.

32 Minutes, WANO Governing Board Meeting, 26 August 1996, 4-7; Minutes, WANO Governing Board Meeting, 6 November 1995, 4.

33 WANO, Biennial Review 1993-1995, 16-17.

34 Ibid., 17.

35 Minutes, WANO Governing Board Meeting, 17 November 1994, Closed Session, 1-2; Minutes, WANO Governing Board Meeting, 23 April 1995, 5; Pozdyshev biography, WANO Biennial General Meeting Programme, 2002, 20.

36 WANO, Annual Review, 1995, 2, 5.

37 Ibid. 2; Clarke Oral History, 11 May 2013, 82-83.

38 Carle to author, 13 March 2014; WANO, Annual Review, 1995, 2.

39 Clarke Oral History, 11 May 2013, 96; Minutes, WANO Governing Board Meeting, 20 July 1995, 17; Biography of Vince J. Madden, http://www.world-nuclear.org/sym/1999/maddbio.htm.

40 Clarke Oral History, 11 May 2013, 99-101.

41 Ibid., 101-102.

42 Chairman's Report, Prague Biennial General Meeting, 12 May 1997, WANO Archives, AC.

43 Ibid., 1-2.

44 Ibid., 2-3, 4.

45 Ibid., 4-5, 11.

46 Ibid., 5-6.

47 Ibid., 6.

48 Carle Oral History, 14 May 2014, 53; Chairman's Report, Prague Biennial General Meeting, 12 May 1997, 7.

49 Ibid., 13, 16-17.

50 Minutes, WANO Governing Board Meeting, 26 August 1996, 3; ibid., 21 November 1996, 5.

51 Pate Oral History, 4 March 2013, 30-32; author's conversation with Zack Pate, 14 April 2014; *Inside WANO*, No. 16 (August 1997), 3. Toukhvetov replaced Anatoly Kontsevoy, who had served six years as director of Moscow Centre.

52 Carle Oral History, 14 May 2013, 51; Pate Oral History, 4 March 2013, 30-31; Atlanta Centre Governing Board, Meeting Minutes, 11 May 1997.

53 *Inside WANO*, No. 16 (August 1997), 11; Leclercq and Carle Oral History, 14 May 2013, 32-33.

4 SECURING THE MANTLE OF NUCLEAR SAFETY

1 Cantelon and Walker, *Core of Excellence*, 77-78.

2 Ibid., 78-79, 14.

3 Ibid., 79-80.

4 Ibid., 108-109; 93-96.

5 Zack Pate Oral History, 24 January 2013, 6.

6 Ibid., 6-7; ibid., 4 March 2013, 50.

7 Pate Oral History, 24 January 2013, 22-23.

8 Ibid., 23-24.

9 Ibid., 24-25.

10 Allan Kupcis Oral History, 19 April 2013, 1-5; Pate Oral History, 5 March 2013, 11.

11 O. Allan Kupcis, Remarks at Prague BGM, 13 May 1997, *passim*. Kupcis Oral History, 19 April 2013, 8.

12 Pate Oral History, 5 March 2013, 11-12.

13 WANO Chronology, 11-13 May 1997; Meeting Minutes, WANO Governing Board, 11 May 1997 and 22-23 July 1997.

14　WANO Chronology, April 1998; Ignatenko was killed in an automobile accident shortly after his appointment.

15　Pate Oral History, 4 March 2013, 32.

16　Minutes of WANO Governing Board, Meeting No. 27, 30 September and 1 October 1997, 5-6; Minutes of WANO Governing Board, Meeting No. 28, 10 November 1997, 3.

17　Ibid., 7-8; George Felgate to author, 25 September 2014.

18　Minutes of WANO Governing Board, Meeting No. 28, 10 November 1997, 9.

19　Ibid., 8-9; 3- 6; WANO, *Biennial Review*, 1997-1999, 19.

20　WANO Governing Board, Meeting No. 25, 11 May 1997, 3-4; ibid., Meeting No. 27, 30 September and 1 October 1997, 8; ibid., Meeting No. 28, 10 November 1997, 8.

21　Kupcis Oral History, 19 April 2013, 10-12; WANO Governing Board, Meeting No. 27, Closed Session, 30 September and 1 October 1997, 2-3.

22　Pate Oral History, 5 March 2013, 25-27.

23　Minutes of WANO Governing Board Meeting No. 26, 22-23 July 1997, 7; Report of Sig Berg, 21 October 1997, Chernobyl Notebook, WANO Atlanta Centre archives; Pate Oral History, 5 March 2013, 26-29; Kupcis Oral History, 19 April 2013, 23; Zack T. Pate to J. Michael Evans, 23 April 2002, Attachment: WANO Progress Report, 1-2; Zack T. Pate to Nikolay Friedman, 7 August 1997, Attachment to WANO Chairman's Letter to Ukraine of 7 August 1997, 1-4, WANO History Files, Pate Correspondence.

24　Minutes of WANO Governing Board Meeting, No. 26, 22-23 July 1997, 7-8; Pate Oral History, 5 March 2013, 26-29; Willy De Roevere to the European Commission, 10 September 1997, Chernobyl Notebook, WANO Atlanta Centre archives; Pate, WANO Progress Report, 2. As a result of pressure from WANO and INPO, US Vice President Al Gore sent a letter to Ukrainian President Leonid Kuchma expressing concern over the dangers of the Chernobyl and traveled to Ukraine several months later to follow up. Unit 3 was put in an extended outage to correct many of the problems, then was shuttered.

25　Pate to Friedman, 7 August 1997, WANO History Files, Pate Correspondence.

26　Ibid.

27　Kupcis Oral History, 19 April 2013, 23; Minutes of WANO Governing Board Meeting, No. 28, 10 November 1997, 6-7; Pate to J. Michael Evans, 23 April 2002, Attachment: WANO Progress Report, 3, WANO History Files, Pate Correspondence.

28　Pate Oral History, 5 March 2013, 31-32.

29　WANO, *Biennial Review*, 1997-1999, 5; *Inside WANO*, Issue 21 (February 2000), 9; WANO Governing Board, Minutes, Meeting No. 33, 3-4 March 1999, 7; WANO, *Biennial Review*, 1999-2000, 4. Unlike Lee, McRae, and Kupcis, Choi was not an activist WANO President. His brief attendance at many WANO Governing Board meetings, or non-attendance altogether, suggest that he saw his role as a planner for the next BGM rather than an ambassador of WANO. See attendance sheets for WANO Governing Board Meetings in 1999, 2000, and 2001.

30　WANO,*Biennial Review*, 1997-1999, 4, 26-27.

31　Ibid., 5.

32　WANO did Y2K readiness reviews at three Russian and two Ukrainian plants in 1999 to test for potential computer problems. Those reviews served as models elsewhere. No nuclear safety-related issue occurred at any nuclear power plant when the date rolled over to 2000. See WANO, *Biennial Review*, 1999-2000, 10-11, 13.

33　*Inside WANO*, Issue 21 (February 2000), 7.

34　WANO, *Biennial Review*, 1997-1999, 6.

35　WANO Governing Board, Meeting Minutes, 20-21 July 2000, 12-14.

36　Ibid., 14-15 November 2000, Attachment 1, 18.

37　Ibid., 19.

38　Ibid. See also WANO Governing Board, Meeting Minutes, 14-15 November 2000, 6-8.

39　Cape Town Strategy Session, 15-16 November 2000, 1-6.

40　WANO Governing Board, Meeting Minutes, 14-15 November 2000, 13.

41 Sigval M. Berg Oral History, 19 March 2013, 16-20.

42 Pate Oral History, 5 March 2013, 41-43; WANO Chronology, 13-14 June 2001.

43 WANO Governing Board Minutes, 28-29 October 2001, 7.

44 WANO Governing Board, Meeting Minutes, 28-29 October 2001, 2-3, 5; WANO, Review, 2003, 9-13.

45 WANO Governing Board, Meeting Minutes, 28-29 October 2001, 7; ibid. 20-21 February 2002, 6.

46 WANO, Review, 2003, 15; WANO Governing Board, Meeting Minutes, 20-21 February 2002, 3-4.

47 Minutes, WANO Governing Board, Executive Session, 28-29 October 2001, 12; ibid., 20-21

February 2002, 2; 17 March 2002, 2; http://www.zoominfo.com/p/Hajimu-Maeda/240267525; Berg Oral History, 19 March 2013, 13; Hajimu Maeda Oral History, 22 April 2013, 2-6. In all, the WANO Governing Board considered eight different candidates in its search for Pate's successor. See Zack T. Pate, "WANO Chairmanship," February 2004, Pate Correspondence File.

48 Zack T. Pate to J. Michael Evans, 23 April 2002, Attachment: WANO Progress Report, 23 April 2002, 4-8; Felgate to author, 22 September 2014.

49 Minutes, WANO Governing Board, 28-29 October 2001, 3.

50 Zack T. Pate to J. Michael Evans, 23 April 2002, Attachment: WANO Progress Report, 23 April 2002, 8.

5 THE TRIALS OF CHANGE

1 Hajimu Maeda Oral History, 22 April 2013, 3, 7-8; Maeda to author, 23 March 2015.

2 Maeda Oral History, 22 April 2013, 9-10; Maeda to author, 23 March 2015; WANO Governing Board, Meeting Minutes, Executive Session, 17 March 2002, 2. The nominee for WANO president in 2002 was Pierre Carlier.

3 Biography in Safety Culture: Building It, Keeping It, 2002 INPO CEO Conference, 4; Pierre Carlier Oral History, 15 May 2013, 3.

4 Sigval M. Berg Oral History, 19 March 2013, 2-3; Pate Oral History, 5 March 2013, 51-52.

5 Sigval M. Berg Oral History, 19 March 2013, 2-3; Pate Oral History, 5 March 2013, 51-52; William Cavanaugh III Oral History, 26 March 2013, 6; WANO Governing Board, Meeting Minutes, Executive Session, 28-29 October 2001, 12; ibid., 24 April 2003, 2-3. See also Pate Oral History, 4 March 2013, 37-38. Krško moved from Atlanta to the Paris Centre in the 1990s and Koeberg moved from Paris to the Atlanta Centre. See Hutcherson Oral History, 23-24 April 2013, 141.

6 Berg Oral History, 2 November 2010, 2-4, INPO History Archives, Atlanta.

7 Berg Oral History, 19 March 2013, 7.

8 Berg Oral History, 19 March 2013, 8, 13; Maeda Oral History, 22 April 2013, 11-12, 22; Maeda to author, 23 March 2015.

9 Berg Oral History, 19 March 2013, 8, 13; David Igyarto Oral History, 5 March 2013, 4; Berg Oral History, 19 March 2013, 78.

10 WANO 2003 Proposals, "Moving Forward Together," Minutes, Meeting of the WANO-Atlanta Centre Governing Board, 28 May 2003, 2-3 and Attachment, passim.

11 Ibid., 64.

12 Carlier Oral History, 15 May 2013, 3.

13 WANO Governing Board, Minutes, 8-9 October 2002, 5; George Felgate to author, 22 February 2015.

14 Berg Oral History, 19 March 2013, 64-65.

15 George Hutcherson Oral History, 23-24 April 2013, 111.

16 Carlier Oral History, 15 May 2013, 3.

17 WANO Governing Board, Minutes, 8-9 October 2002, 6; ibid., 24 April 2003, 4; ibid., 12 October 2003, 2; Berg Oral History, 19 March 2013, 66-68.

18 Berg Oral History, 19 March 2013, 66-68; Maeda Oral History, 22 April 2013, 15; Ed Hux Oral History, 21 March 2013, 12-13; Felgate to author, 22 February 2015.

19 Berg Oral History, 68; Maeda to author, 23 March 2015.

20 http://en.wikipedia.org/wiki/Paks_Nuclear_Power_Plant#2003_incident; "The Future of Nuclear Power in Central and Eastern Europe," Conference Proceedings, Budapest, 19 October 2003.

21 Maeda Oral History, 22 April 2013, 15; Maeda to author, 23 March 2015.

22 James Varney, "Good Is Not Good Enough," *Modern Power Systems,* 9 September 2004 [http://www.modernpowersyastems.com/story.asp?sc=2024453]. Report of WANO Self-Assessment of the Tokyo Centre, 12 June 2003, 1, WANO History Archives. All the regional centres, including the London Coordinating Centre, underwent self-assessments between 2003 and 2004.

23 Varney, "Good Is Not Good Enough," *Modern Power Systems,* 9 September 2004. [http://www.modernpowersyastems.com/story.asp?sc=2024453]

24 Maeda Oral History, 24 April 2003, 2-3; Hux Oral History, 21 March 2013, 16-17; Felgate to author, 22 February 2015; WANO Governing Board, Minutes, 24 April 2003, 3.

25 WANO Governing Board, Minutes, 24 April 2003, 3.

26 Ibid., 6; Carlier Oral History, 15 May 2013, 10; Maeda Oral History, 22 April 2013, 24-25.

27 Report of WANO Self-Assessment of the WANO Self-Assessment of the Tokyo Centre, 12 June 2003, 1. The methodology was the same for all centres.

28 Report of the WANO Self-Assessment of the Atlanta Centre, 16 April 2003, 2, 5, 7-8; Hutcherson Oral History, 23-24 April 2013, 102.

29 Ibid., 102; Report of WANO Self-Assessment

of the Tokyo Centre, 12 June 2003, 2-18.

30 Felgate to author, 22 February 2015.

31 Report of WANO Self-Assessment of the Moscow Centre, 30 June 2003, 2-15. The Moscow Centre's website was in Russian. Its creation and use meant that the English language WANO website was seldom used by Moscow Centre members. Felgate to author, 22 February 2015.

32 Report of WANO Self-Assessment of the Moscow Centre, 30 June 2003, 2-15; Felgate to author, 22 February 2015.

33 Igyarto Oral History, 5 March 2013, 8-10; Felgate to author, 22 February 2015.

34 Report of WANO Self-Assessment of the Paris Centre, 13 January 2004, 2-20; Carlier Oral History, 15 May 2013, 10.

35 Minutes, Executive Session of WANO Governing Board, 16 January 2004, 2; Abagyan quoted in Minutes, WANO Governing Board, 13-14 October 2004, 9.

36 Report of WANO Self-Assessment of the Coordinating Centre, 9 March 2004, 3-17; Felgate to author, 22 February 2015.

37 Ibid.

38 WANO Governing Board, Minutes, 6-7 April 2004, 3; Berg Oral History, 19 March 2013, 78-79; Igyarto Oral History, 5 March 2013, 5.

39 WANO Governing Board, Minutes, Executive Session, 16 January 2004, 2-4, 12.

40 Felgate to author, 22 February 2015.

41 WANO Governing Board, Minutes, Executive Session, 16 January 2004, 2-4, 12.

42 Ibid., 4-5.

43 Maeda Oral History, 22 April 2013, 26-27. Saraev spoke better English than he acknowledged publicly, according to Berg. See Berg Oral History, 19 March 2013, 75.

44 Minutes, WANO Governing Board, Executive Session, 23 July 2004, 2; William Cavanaugh III Oral History, 26 March 2013, 46.

45 Maeda to author, 23 March 2015.

46 William Cavanaugh III Oral History, 22 November 2010, 1-2, 8, 19-20, INPO Archives, Atlanta; Cavanaugh Oral History, 26 March 2013, 2-4.

47 Ibid.

48 Lucas Mampaey Oral History, 9 April 2013, 2-5, 12, 14.

49 Ibid.

50 Ibid., 15; Cantelon and Walker, *Core of Excellence*, 167.

51 Mampaey Oral History, 9 April 2013, 16-17; Cavanaugh Oral History, 26 March 2013, 21-22.

52 See Ibid., 9-12. Minutes, WANO Governing Board, 13-14 October 2004, 5. On the nuclear renaissance, see William Tucker, "Nuclear Proliferation: An Industry Rises from the Dead," *The Weekly Standard*, 6 June 2006.

53 Hutcherson Oral History, 23-24 April 2013, 137-138, 103.

54 "WANO Long-Term Plan," Attachment to Minutes, Executive Session of WANO Governing Board, 6-7 April 2005, 3-5.

55 Minutes, WANO Governing Board, 6-7 April 2005, 4; ibid., 9 October 2005, 6; ibid., 12-13 April 2006, 2. In 2004 Moscow Centre had conducted a peer review of a nuclear-powered icebreaker, the *Vaigach*. See Minutes, WANO Governing Board, 13-14 October 2004, 3; Felgate to author, 22 February 2015.

56 Minutes, Atlanta Centre Governing Board, 24 May 2005, 3; Minutes, WANO Governing Board, 6-7 April 2005, 4-5; ibid., 20-21 July 2005, 5, 3.

57 Ibid., 5; Minutes, WANO Governing Board, 13-14 October 2004, 5, 8. The Japanese had formed a new organisation, the Japan Nuclear Technology Institute (JANTI), somewhat along the lines of INPO, with hopes that it would provide trained and experienced peer reviewers with an "international perspective."

58 See Minutes, WANO Governing Board, 6-7 April 2005, 2.

59 Dick Kovan, "WANO Seeks Renewal," *Nuclear News*, December 2005, 49-52.

60 Ibid. See also Hawthorne to WANO-Atlanta Centre Governing Board, 16 May 2006, Attachment 2.

61 Minutes, WANO Governing Board, Executive Session, 12-13 April 2006, 3; Felgate to author, 22 February 2015.

62 Oliver D. Kingsley, Jr., to Duncan Hawthorne, Aleš John, Stane Rožman, and Jianfeng Yu, 23 May 2006, Special Committee, 2006 Folder, WANO Archives. Cavanaugh was re-elected at the Governing Board meeting in Beijing, People's Republic of China, in April 2006.

63 Duncan Hawthorne to David Hay and Thierry Vandal, 4 August 2006, WANO Special Committee Folder, Atlanta Centre; Ann MacLachlan, "WANO Warns Safety Lapse Anywhere Could Halt 'Nuclear Renaissance,'" *Nucleonics Week*, 27 September 2007 (Vol. 48, No. 39), l. EDF was a prime example of the growing internationalisation of nuclear power. The company formed a transatlantic partnership in late 2008 with Baltimore-based Constellation Energy and bought British Energy in 2009.

64 Minutes, WANO Governing Board, 12-13 July 2006, 6; Roger Spinnato Oral History, 24 January 2013, 26-29.

65 Minutes, WANO Governing Board, 12-13 July 2006, 4-5; S.K. Jain to Duncan Hawthorne, 14 June 2006, WANO Special Committee Folder; Jianfeng Yu to Kingsley, 5 July 2006, WANO Special Committee Folder.

66 Minutes, Executive Session, WANO Governing Board, 18-19 October 2006, 3-4; Minutes, Governing Board Meeting, Atlanta Centre, 31 October 2006, 4.

67 Spinnato Oral History, 24 January 2013; Minutes, Executive Session, WANO Governing Board, 18-19 October 2006, 2; Minutes, Governing Board Meeting, Atlanta Centre, 23 September 2007, 3.

68 Minutes, WANO Governing Board, 18-19 April 2007, 6; ibid., 11-12 July 2007, 5; Felgate to author, 22 February 2015; Cavanaugh quoted in WANO Chronology, 34.

69 Ibid., 15 January 2009, 5-6, 10-11.

6 LAST CHANCE TO GET IT RIGHT

1 Carlier quoted in Minutes, WANO Governing Board, 29-30 July 2009, 12.

2 Minutes, WANO Governing Board, Executive Session, 6-7 August 2008, 7.

3 Laurent Stricker Bio, USNRC, http://nrc.gov/public-involve/conference-symposium/ric/past/2012; *Les Echos*, 21 January 2009; Laurent Stricker Oral History, 14 May 2013, 1-3.

4 Ibid., 6-10.

5 Ibid., 11. See Cantelon and Walker, *Core of Excellence*, 199-200. The EDF-Constellation deal later fell through.

6 Minutes, WANO Governing Board, Executive Session, 12-13 October 2006, 2; ibid., 12-13 July 2007, 2-3; ibid., 9 April 2008, 2; http://de.wikipedia.org/wiki/Walter_Hohlefelder; Minutes, WANO Governing Board, Executive Session, 9 April 2008, 3-4.

7 Minutes, WANO Governing Board, Executive Session, 9 April 2008, 3-4.

8 Ibid., 2-5.

9 Ibid., 15 January 2009, 2.

10 Minutes, WANO Governing Board, Executive Session, 6-7 August 2008, 2-3.

11 Carlier Oral History, 15 May 2013, 17-21; Stricker Oral History, 14 May 2013, 12. Stricker received a majority vote of the paper ballots cast. The other candidate, proposed by Atlanta Centre, was Duncan Hawthorne. See Minutes, WANO Governing Board, Executive Session, 16 January 2009.

12 George Felgate Oral History, 24 September 2013, 1-17; Felgate to author, 22 February 2015.

13 Ibid.

14 WANO, "2009 Year-End Highlights Report," n.p; Felgate to author, 22 February 2015.

15 WANO, "2009 Year-End Highlights Report," n.p.; "Two Decades of WANO," *World Nuclear News*, 19 May 2009; see also Anthony Faiola, "Nuclear power regains support," *Washington Post*, 24 November 2009, on the promise of nuclear energy to reduce greenhouse gas emissions.

16 Brian Schimmoller, "Nuclear Performance: A Shifting Target, *Power Engineering*, September 2010, 16.

17 WANO, 2010 Year-End Report, January 2011, 11; *Inside WANO*, vol. 18, no. 1 (2010), 4.

18 "Ensuring Excellence at the Plants," *Nuclear Plant Journal*, January-February 2010, 20, 22.

19 George Felgate, "Positioning WANO for Future Success," *Inside WANO*, vol. 18, no. 1 (2010), 6; Felgate to author, 22 February 2015.

20 Ibid. The regional directors working with Felgate on the ELT were David Farr, Mikhail Chudakov, Ignacio Araluce, and Hal Shirayanagi.

21 *Inside WANO*, vol. 18, no. 1, 4 (2010), 3.

22 WANO, "2009 Year-End Highlights Report," n.p; *Inside WANO*, vol. 18, no. 1, 8.

23 Felgate to author, 22 February 2015; WANO, "2009 Year-End Highlights Report," N.P.; WANO, 2010 Year-End Report, January 2011, 38; "Changes for International Nuclear Safety," *World Nuclear News*, 28 April 2010; *Inside WANO*, vol. 18, no. 1, 8; Felgate to author, 22 February 2015. After Fukushima Tokyo Centre had no effective "big dog," according to Felgate, as TEPCO and other Japanese utilities looked inward, dealing with their own problems.

24 WANO, "2009 Year-End Highlights Report," n.p.; Minutes, WANO Governing Board, Executive Session, 31 July 2010, 2.

25 WANO, 2010 Year-End Report, January 2011, 38-39.

26 Ibid., 39.

27 Felgate to author, 22 February 2015.

28 WANO, 2010 Year-End Report, January 2011, 39-40, 6.

29 Ibid., 9-23.

30 Ibid., 23-40.

31 Ibid., 41; Felgate to author, 22 February 2015.

32 WANO, 2010 Year-End Report, January 2011, 41.

33 Cantelon and Walker, *Core Of Excellence*, 209.

34 Ibid.

35 Ibid., 210.

36 George Felgate Oral History, 24 September 2013, 25-26.

37 Ibid., 26-27; Cantelon and Walker, *Core Of Excellence*, 211.

38 Felgate Oral History, 24 September 2013, 27; Felgate to author, 22 February 2015.

39 Cantelon and Walker, *Core of Excellence*, 6-9.

40 Ibid., 180, 200-202; James O Ellis to author, 5 June 2015.

41 Minutes, WANO Governing Board, Executive Session, 30 March 2011, 5.

42 Cantelon and Walker, *Core of Excellence*, 202; Felgate Oral History, 24 September 2013, 28-30.

43 Minutes, WANO Governing Board, Executive Session, 30 March 2011, 2; Felgate to author, 22 February 2015; author's conversation with Felgate, 4 March 2015.

44 Minutes, WANO Governing Board, Executive Session, 30 March 2011, 3.

45 Ibid., 3-6.

46 Ibid., 3-6.

47 Ibid., 4.

48 http://www.pbnc2014.org/media/uploads/bios/Mitchell%20BIO%202012%20Sept%20PDF.pdf

49 WANO, 2011 Report, 53, 3; WANO Governing Board Meeting, Executive Session, 13 July 2011, 5; WANO, 2011 Report, 9.

50 Ibid., 6.

51 Brian Schimmoller, "Will the New WANO Have Enough Teeth?" *Power Engineering*, January 2012, 14.

52 Ibid., 6-8; *World Nuclear News*, 28 October 2011; Schimmoller, "Will the New WANO Have Enough Teeth?" *Power Engineering*, January 2012, 6-8.

53 Ibid.

54 Ibid., 8-10; WANO Post-Fukushima Commission, Final Report, 30 September 2011, 3.

55 WANO, 2011 Report, 9; WANO Governing Board Meeting, Executive Session, 13 July 2011, 5; WANO Post-Fukushima Commission, Final Report, 30 September 2011, 7-8.

56 *Inside WANO*, vol. 18, no. 1 (2010), 4; Felgate Oral History, 24 September 2013, 55.

57 WANO Post-Fukushima Commission, Final Report, 30 September 2011, 9-10.

58 Ibid., 10-11.

59 Ibid., 11-13.

60 Ibid., 2.

61 Ibid., 13-14; WANO, 2011 Report, 9.

62 WANO Post-Fukushima Commission, Final Report, 30 September 2011, Attachment 3; Felgate to author, 22 February 2015.

63 WANO, 2011 Annual Report, 39, 42; "Moves to Strengthen WANO," *World Nuclear News*, 28 October 2011; *Inside WANO*, vol. 19, no. 3 (2011), 4, 6.

64 "Moves to Strengthen WANO," *World Nuclear News*, 28 October 2011.

65 Brian Schimmoller, "Will the New WANO Have Enough Teeth?" *Power Engineering*, January 2012, 14.

7 ONE WANO

1 Pierre Gadonneix, "Strengthen Nuclear Safety," *Business Daily Update*. 27 March 2012.

2 "Stronger Ties Join IAEA and WANO," *World Nuclear News*, 18 September 2012.

3 See Will Dalrymple, "What Happened Next at the World's Association," *Nuclear Engineering*, 2 August 2013.

4 WANO Post-Fukushima Commission, Final

Report, 30 September 2011, 3, 14.

5 George Felgate to author, 22 February 2015; WANO Internal Assessment Summary Report [Draft], 7 January 2013, 1.

6 Ibid., 1, 6.

7 Pate to Mitchell, 7 March 2012, WANO Archives, Pate Folder.

8 WANO Internal Assessment Summary Report [Draft], 7 January 2013, 6-7.

9 Ibid., 8-9.

10 Ibid., 9-12.

11 Ibid., 13.

12 Minutes, WANO Governing Board, 31 January 2012, 5; ibid., 15 May 2012, 7; Felgate to author, 22 February 2015. A previous attempt to locate the prestart-up office in Shenzhen had failed due to legal issues.

13 *The Asahi Shimbun*, 16 March 2012: http://ajw.asahi.com/article/0311disaster/fukushima/AJ201203160064; Minutes, WANO Governing Board, 15 May 2012, 2; ibid., 5 July 2012, 3, 5.

14 Ibid., 6; ibid., 29 January 2013, 4; Felgate to author, 9 May 2015.

15 Ibid., 31 October 2012, 3; http://www.genanshin.jp/english/association/establishment.html; Minutes, WANO Governing Board, 29 January 2013, 6.

16 http://www.genanshin.jp/english/association/establishment.html; Minutes, WANO Governing Board, 25 July 2012, 6.

17 Minutes, WANO Governing Board, 25 July 2012, 6-8; ibid., 31 October, 2012, 2, 6.

18 Ibid., 29 January 2013, 3-4.

19 Ibid., 19 May 2013, 2-3; S. K. Jain Oral History, 19 May 2013.

20 Minutes, WANO Governing Board, 31 October 2012, 4- 5.

21 Felgate Oral History, 24 September 2013, 85-86; Pate to Mitchell, 7 March 2012, WANO Archives, Pate Folder.

22 Felgate Oral History, 24 September 2013, 85-86; Stricker, "One Year on from Shenzhen," *Inside WANO*, vol. 20, no. 3 (2012), 3; Stricker quoted in Declan Butler, "Nuclear Safety Chief Calls for Reform," *Nature*, 472 (2011)

274; Felgate to author, 22 February 2015.

23 Robert Willard Oral History, 13 October 2015, 14-16; Jacques Régaldo Oral History, 23 February 2015, 10; WANO Governing Board Minutes, Executive Session, 31 October 2012, 5.

24 Régaldo Oral History, 23 February 2015, 1-3; "Jacques Régaldo Appointed Chairman of the Board," WANO Press Release, 31 October 2012; Régaldo to author, 30 March 2015.

25 Régaldo Oral History, 23 February 2015, 19-21, 3.

26 WANO Governing Board, Open Session, 29 January 2013, 2; Kenneth Ellis Oral History, 20 February 2015, 8.

27 Kenneth Ellis Oral History, 20 February 2015, 1-4, 7.

28 Ibid., 4-5, 7.

29 Ibid., 7-10.

30 *Inside WANO*, Vol. 21, No. 2 (2013), 3-4.

31 "Introducing the New WANO President," *Inside WANO*, Vol. 21, No. 2 (2013), 9.

32 Ibid., 5. Before the Chairman's and Managing Director's initial terms expired in 2015, the Governing Board moved to elect Régaldo to an additional two- year term as Chairman, to be considered the equivalent to the recently authorised under the WANO Articles of Association. At the same time, Ellis was reappointed Managing Director for an additional one-year term, meeting the three-year term under the revised Articles of Association. Each would be eligible for an additional term. See WANO Governing Board, Meeting Minutes, Closed Session, 4 June 2014, 2.

33 *Inside WANO*, Vol. 21, no. 2 (2013), 12-15; Minutes, WANO Governing Board, 11 September 2013, 4; "Welcome to the New Face of WANO," Press Release, 12 May 2014. The lack of staff in the London Office in 2012 had produced a £2 million surplus that was returned to members by crediting their dues against their 2014 fees. WANO Governing Board, Meeting Minutes, 11 September 2013, 4.

34 Régaldo proposed the change to the chief executive officer in December 2013. See Minutes, WANO Governing Board, 3 December 2013, 7.

35 WANO Governing Board, Meeting Minutes,

Open Session, 19 May 2013, 3; *Inside WANO*. Vol. 21, no. 2 (2013), 10; ibid., Vol. 21, no. 3 (2013), 12-14.

36 Governing Board, Meeting Minutes, Executive Session, 19 May 2013, 2-4; Governing Board, Meeting Minutes, Closed Session, 11 September 2013, 3.

37 Governing Board, Meeting Minutes, Open Session, 3 December 2013, 2; *Inside WANO*, Vo. 22, no. 3 (2014), n.p.

38 Governing Board, Meeting Minutes, Open Session, 4 December 2014, 6.

39 Governing Board, Meeting Minutes, Open Session, 11 September 2013, 2-3.

40 Ibid., 3; Governing Board, Meeting Minutes, Open Session, 4 March 2014, 4; Kenneth Ellis to author, 7 May 2015.

41 Ibid., 2-3; Kenneth Ellis to author, 7 May 2015.

42 Ellis Oral History, 20 February 2015, 59-60.

43 *Inside WANO*, Vol. 22, No. 3(2014), n.p.; WANO Governing Board, Meeting Minutes, Open Session, 4 September 2014, 8; Ellis Oral History, 20 February 2015, 59-60.

44 WANO Governing Board, Meeting Minutes, Closed Session, 4 June 2014, 5; WANO Governing Board, Meeting Minutes, Open Session, 4 September 2014, 2; Régaldo Oral History 23 February 2015, 22-24.

45 WANO Governing Board, Meeting Minutes, Open Session, 4 June 2014, 6.

46 Ibid., 4-5; Kenneth Ellis to author, 7 May 2015.

47 "Strengthening Station Performance," *Inside WANO*, Vol. 22, No. 4 (2014), n.p.

48 WANO Governing Board, Meeting Minutes, Open Session, 4 September 2014, 3-4; WANO Governing Board, Meeting Minutes, Restricted Session, 4 September 2014, 4.

49 *Compass*, WANO Long-Term Plan 2015-2019, 4-5.

50 Ibid., 5-9, 17-25.

51 WANO Policy Document 10, Plant of Focus, adopted by WANO Governing Board, April 2015, 22. Initially, there was considerable resistance to the idea of plant rankings and highlighting and naming plants that were troubling or high risk. Importantly, according to one senior WANO executive, "We have been able to overcome both of these [issues] in the past couple of years under Ellis' leadership". See Robert Willard Oral History, 13 October 2015, 8.

52 Ibid., 26, cover.

53 Duncan Hawthorne Oral History, 30 September 2015, 4-6.

54 Thomas N Mitchell Oral History, 6 October 2015, 26, 12, 7-8.

55 Ibid. 12, 14-15.

56 Kenneth Ellis, WANO Update, 5 October 2015, passim.

57 Ibid; Jacques Régaldo, Closing Remarks, "One WANO," Toronto BGM, 5 October 2015, 5;

58 Mitchell Oral History, 6 October 2015, 36, 30-32, 25-27, 10.

59 Robert Willard Oral History, 13 October 2015, 3-4.

60 Hawthorne Oral History, 30 September 2015, 7, 11-12.

61 Jacques Régaldo, Opening Remarks, "One WANO," Toronto BGM, 5 October 2015, 8.

62 Jacques Régaldo, Opening Remarks, "One WANO," Toronto BGM, 5 October 2015, 9-10; Mitchell Oral History, 6 October 2015, 40-41.

63 Régaldo Oral History, 23 February 2015, 69, 47-48, 60, 67.

64 Jacques Régaldo, Closing Remarks, "One WANO," Toronto BGM, 6 October 2015, 1-4.

65 Mitchell Oral History, 6 October 2015, 36; Hawthorne Oral History, 30 September 2015, 15. Over its history WANO has adopted regional ideas and programmes and implemented them on an international level.

66 Jacques Régaldo, Closing Remarks, "One WANO," Toronto BGM, 6 October 2015, 3, 6-7.

67 George Felgate to author, 11 May 2015.

WANO OFFICERS

Chairman		President	BGM
		Nikolai Lukonin	Inaugural Meeting (1989)
Walter Marshall	1989–1991	William S "Bill" Lee III	Atlanta (1991)
	1991–1993	Shoh Nasu	Tokyo (1993)
Rémy Carle	1993–1995	Ian McRae	Paris (1995)
	1995–1997	Eric Pozdyshev	Prague (1997)
Zack T Pate	1997–1999	Allan Kupcis	Victoria, BC (1999)
	1999–2001	Soo-byung Choi	Seoul (2002)
	2001–2002	Pierre Carlier	Berlin (2003)
Hajimu Maeda	2002–2004	Oleg Saraev	Budapest (2005)
William Cavanaugh III	2004–2006	Oliver Kingsley	Chicago (2007)
	2006–2009	Shreyans Kumar Jain	New Delhi (2010)
Laurent Stricker	2009–2011	Qian Zhimin / HE Yu	Shenzhen (2011)
	2011–2013	Vladimir Asmolov	Moscow (2013)
Jacques Régaldo	2013–present	Duncan Hawthorne	Toronto (2015)
		CHO Seok	Gyeongju (2017)

LONDON COORDINATING CENTRE / LONDON OFFICE

Thomas Eckered	Director	1989–1992
Walter J Coakley	Acting Director	1992
Andrew Clarke	Director	1992–1995
Vincent Madden	Director	1995–2001
Anthony Capp	Director	2001–2002
Sigval Berg	Managing Director	2002–2004
Lucas Mampaey	Managing Director	2004–2009
George Felgate	Managing Director	2009–2013
Kenneth Ellis	Managing Director	2013–2014
Kenneth Ellis	Chief Executive Officer	2014–2015
Peter Prozesky	Chief Executive Officer	2016-present

WANO NUCLEAR EXCELLENCE AWARD WINNERS

2003–2013

The WANO Board of Governors established the WANO Nuclear Excellence Award in July 2002 "in honour of Dr Zack T Pate and in recognition of his leadership in promoting excellence in the world-wide nuclear industry". Individuals at any level whose work contributes to or supports the successful operation of nuclear power plants operated by any WANO member(s) are eligible to receive the award, which is made by an independent selection committee with at least one representative from each region. Recipients are honoured at WANO Biennial General Meetings.

2003

Rebba Bhiksham
Nuclear Power Corporation of India Limited

Won-yong Chung
Korea Hydro and Nuclear Power Company

Pedro Figueiredo
Eletrobras Termonuclear SA – Eletronuclear

Bernard Fourest
Électricité de France

Oliver D Kingsley, Jr
Exelon Corporation

Paul Spekkens
Ontario Power Generation

2005

Armen Abagyan
OAO VNIIAES and Concern Rosenergoatom

An-Hong Jeng
Taiwan Power Company

Vladimir Korovkin
Rovno Nuclear Power Plant, Energoatom

Keith Moser
Exelon Corporation

Peter Prozesky
British Energy

Alfred C Tollison Jr
Institute of Nuclear Power Operations

2007

Jussi Helske
Fortum Corporation

Dr Michio Ishikawa
Japan Nuclear Technology Institute

Louis B Long
Southern Nuclear Operating Company

Guntur Nageswara Rao
Nuclear Power Corporation of India Limited

Mana K Nazar
American Electric Power

Pierre Wiroth
Électricité de France

2010

Shashi Bhattacharjee
Nuclear Power Corporation of India Limited

Bill Coley
Formerly of British Energy

Rafael Fernandez de la Garza
Comisión Federal de Electricidad

Rhonda A Lightfoot
Bruce Power

Alexander Lokshin
Rosatom State Corporation

Amir Shahkarami
Exelon Generation

Viktor Shevaldin
Ministry of Energy of the Republic of Lithuania

Yunlong Zan
China Guangdong Nuclear Power Holding Co Ltd.

2011

Akram Ahsan
Pakistan Atomic Energy Commission

CHEN Hua
China National Nuclear Corporation

George Hutcherson
Institute of Nuclear Power Operations

Manuel D Ibañez
Spanish Electricity Industry Association (UNESA)

Susan Reilly Landahl
Exelon Corporation

LU Changshen
Daya Bay Nuclear Power Operations and Management Company

Stane Rožman
Nuklearna Elektrarna Krško

Nikolay Sorokin
Concern Rosenergoatom

2013

Fred Dermarkar
Ontario Power Generation

Scot Greenlee
Exelon Generation

Dominique Minière
Électricité de France

Ján Naňo
Slovenske Elektrarne

Vasily Omelchuk
Concern Rosenergoatom

Matt Sykes
EDF Energy

Takao Watanabe
Tohoku Electric Power Company Inc

ZHANG Shanming
China Guangdong Nuclear Power Holding Co Ltd

2015

Zbyněk GRUNDA, ČEZ, a.s.
Dukovany Nuclear Power Plant

Miraj Ahmad Khalid
Chashma Nuclear Power Generation Stations (CN-PGS) Unit 2

Thomas Mitchell
Ontario Power Generation

Gary Newman
Bruce Power

Andrey Petrov
JSC "Concern Rosenergoatom"

Ami Rastas
Teollisuuden Voima Oyj (TVO)

MA Shu
China General Nuclear Power Corporation (CGNPC)

O J "Ike" Zeringue
Tennessee Valley Authority (TVA

WANO ORAL HISTORY INTERVIEWS

Asmolov, Vladimir
Berg, Sigval
Carle, Rémy
Carlier, Pierre
Cavanaugh, William III
Clarke, Andrew
Eckered, Thomas
Ellis, Kenneth
Farr, David
Felgate, George
Hutcherson, George
Hux, Edgar
Igyarto, David
Jain, SK
John, Aleš
Kirichenko, Anatoly
Kupcis, O Allan
Leclercq, Jacques
Maeda, Hajimu
Mampaey, Lucas
Martin, Timothy
McDonald, Graham
Pate, Zack T
Regaldo, Jacques
Spinnato, Roger
Stricker, Laurent
Takekuro, Ichiro

INDEX